THE PIG DAY BOOK

First published in 2003 by
Michael O'Mara Books Limited
9 Lion Yard, Tremadoc Road,
London SW4 7NQ

ISBN: 1-84317-061-2

1 3 5 7 9 10 8 6 4 2

Designed by Design 23

Origination by Colourwise-Kco

Printed and bound by Eurolitho SpA, Milan

THE PIG DAY BOOK

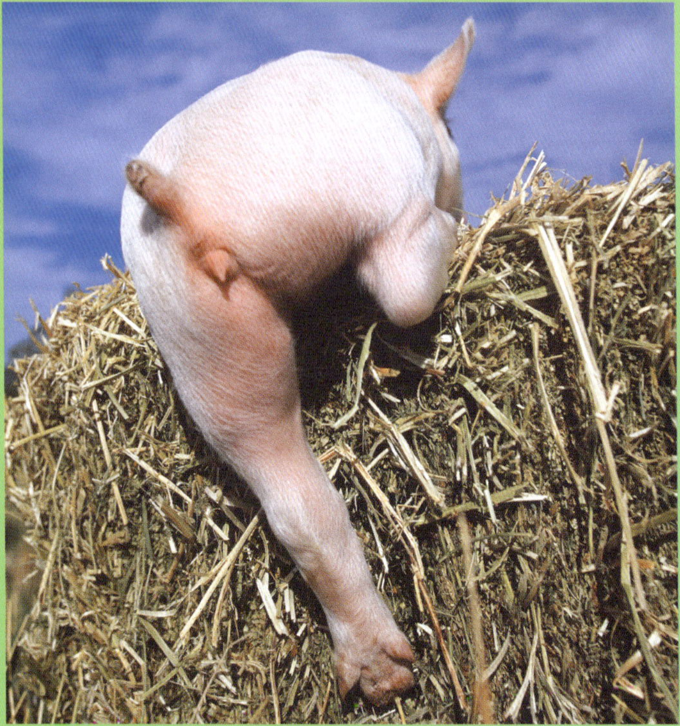

Pig-tail

Percy is determined to see the other side of the haystack.

INTRODUCTION

The pig has a long and distinguished history. It is very rare for an animal species to be found naturally on all continents, except Antarctica – a trotter in every port, you might say! Ranging from the largest, the Giant Forest Hog, through bearded and tusked pigs, to the smallest, the African pygmy pig, there are over 180 breeds of pig worldwide. China was the first country to begin breeding them domestically, in 7,500 BC, and even today is home to the largest population of domestic porkers. Their prolific breeding habits, with an average of fifteen piglets per litter, has always made them popular in agricultural circles, and pigs have certainly made their mark on the world. One of the most famous streets in the world has porcine origins: Wall Street in New York was named after a fence built by the residents of Manhattan to keep rampaging pigs at bay.

The pig is a creature often much maligned, with as many detractors as fans. It has been commonly characterized, along with many other farm animals, such as sheep and cows, as not having much going on between those pink ears. Those in the know, however, have a better opinion of the pig. Scientists have ranked pigs as the fourth most intelligent species in the world, rivalling apes and dolphins for sheer brain power. They are considered by animal experts to be just as trainable and loyal as dogs, and more so than cats! In days gone by, sailors would always carry at least one pig with them on board because they believed that in the event of a shipwreck, a pig would always know to swim to the nearest shore.

The notion that pigs are dirty creatures, purely because they enjoy wallowing around in mud, is also a fallacy: in fact, they do not have sweat glands to help regulate their body temperature, and so the mud provides a vital means of keeping cool. As pigs are very prone to sunburn, mud is also helpful to them by acting as a primitive kind of sunblock!

Today, all over the world, breeding pigs is big business. Pork is the most widely eaten meat across the globe, and in the USA alone almost 60 million pigs are bred each year for food. However, while pigs may be bringing home the bacon for meat-eaters, there are those of us who believe that the best place for a pig is not necessarily in a hot oven with an apple in its mouth! It is becoming increasingly common for people to keep pigs as pets, and those who do find the practice immensely rewarding. Quite apart from being easy to train, they get on well with other domestic animals, and are very partial to a cuddle or a fun game of fetch. The main thing to be wary of is choosing a type of pig that will not grow to too large a size for your house – some people have been caught out with small piglets that grew to double the size they had anticipated; though few have ever had the experience of the owners of Big Bill, the largest-ever hog, who grew to weigh a record-breaking 2,552 lbs!

Pigs have had many famous fans throughout the ages. Such luminaries as the Roman poet Horace, the Scottish novelist and poet Sir Walter Scott and American film star George Clooney have all been avowed porcinophiles. Famously cast as tyrants in George Orwell's *Animal Farm*, the pig's representation in culture is difficult to pin down. From the grande dame of the English country estate in P.G. Wodehouse's 'Blandings' series, through the loveable striplings of Dick King Smith's *Charlotte's Web* and the buffoonish Porky Pig, to the intrepid hero of the film *Babe*, pigs, it seems, can be many things to many different people. This is perhaps a testament to the myriad of myths and legends that have always surrounded the pig, and to the strength and wide appeal of its character.

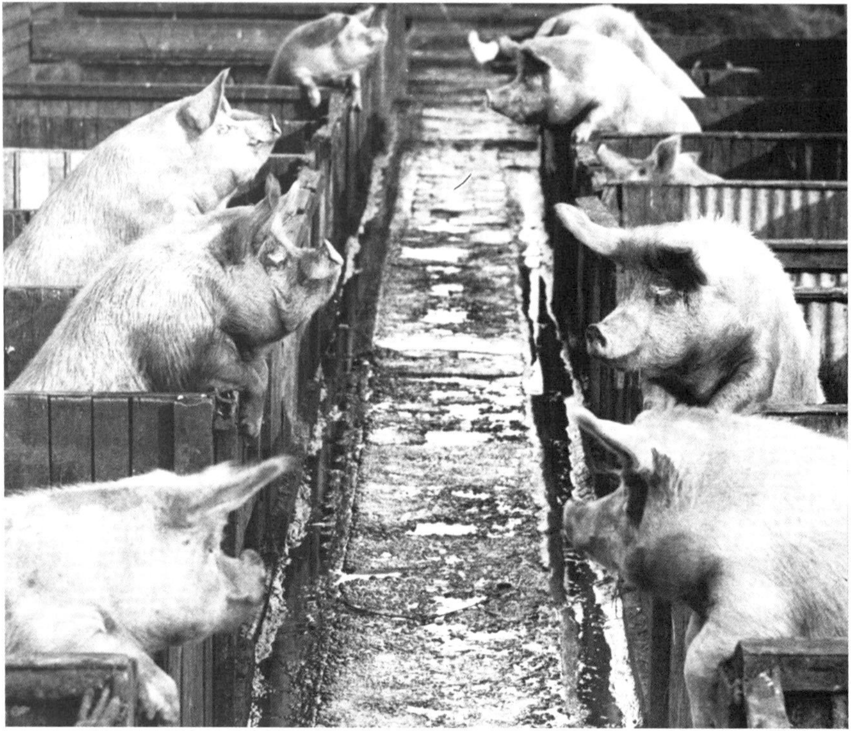

Pork Up Your Ears

The sows down on the farm exchange crackling bits of gossip.

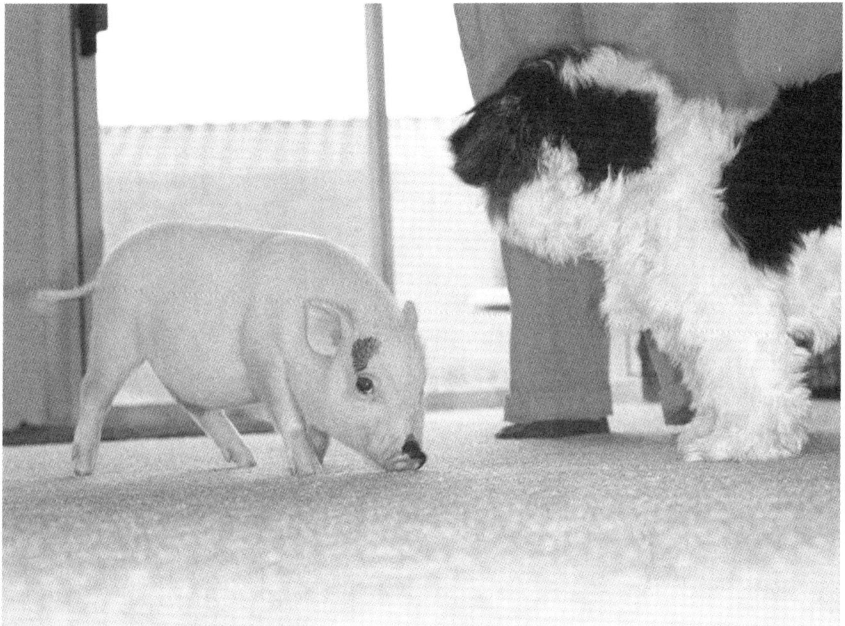

Hoovering Around

Dougal tries to ignore Schnowser vacuuming up tasty morsels.

Belly Laugh

George wonders if Emily has an itch that needs scratching.

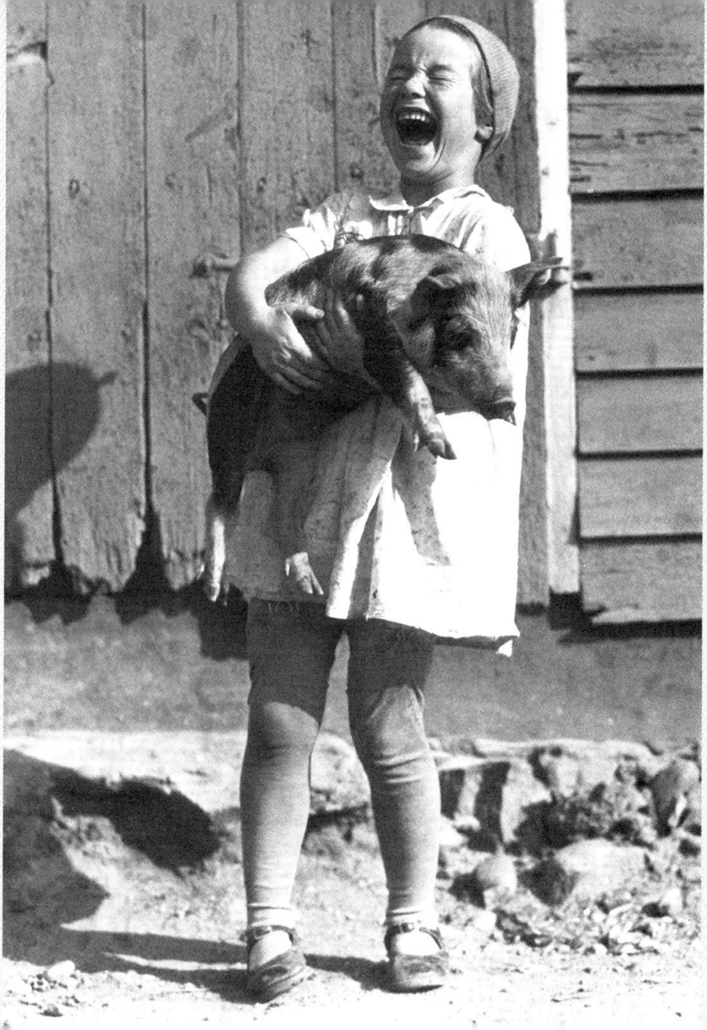

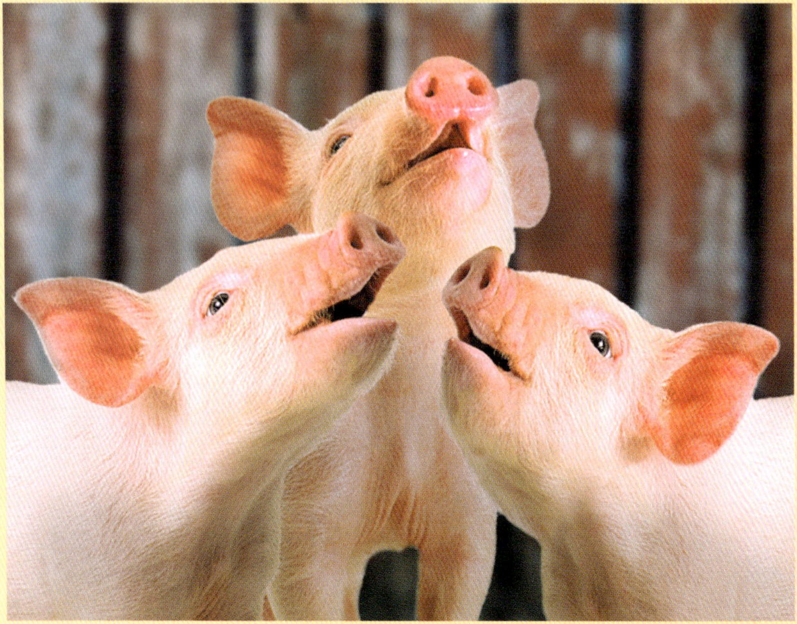

The Three-piggy Opera

The singing sisters are clearly in ham-ony with each other.

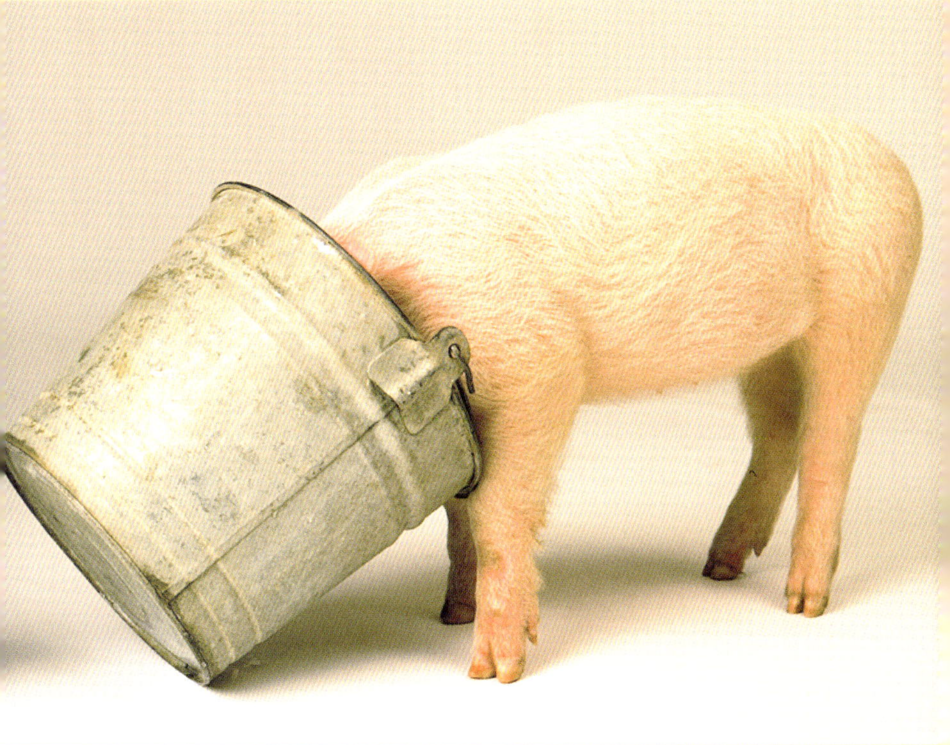

In the Can

Pedro prefers country music to the classics.

Sleeping Beauty

Sadie dreams of making a silk purse...

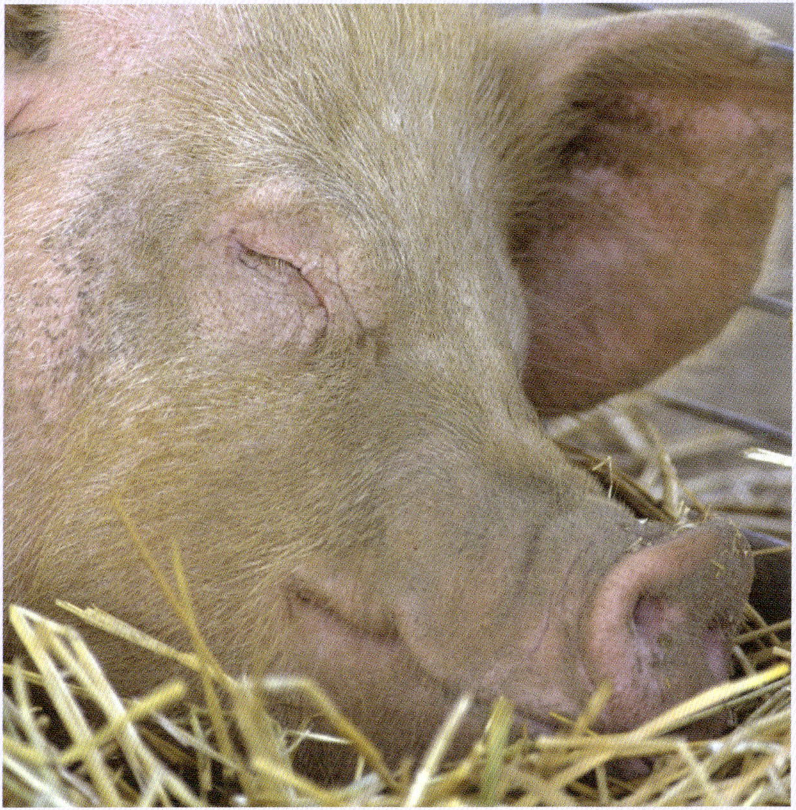

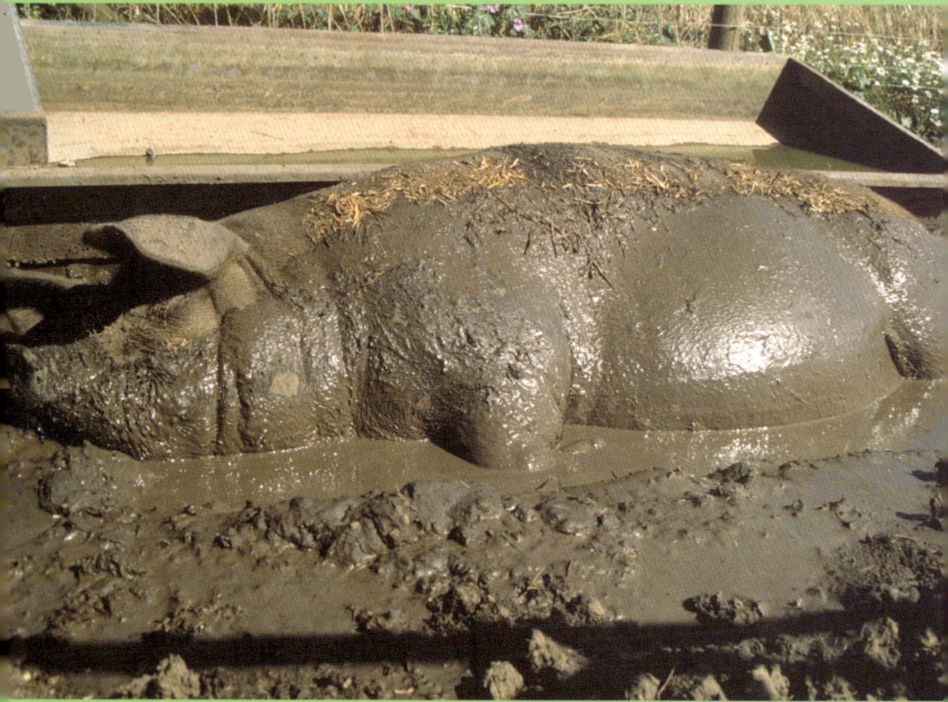

Beauty is Skin-Deep

Molly takes a beauty treatment down at the spa.

Cooling-Off

After the mud comes the hot-tub.

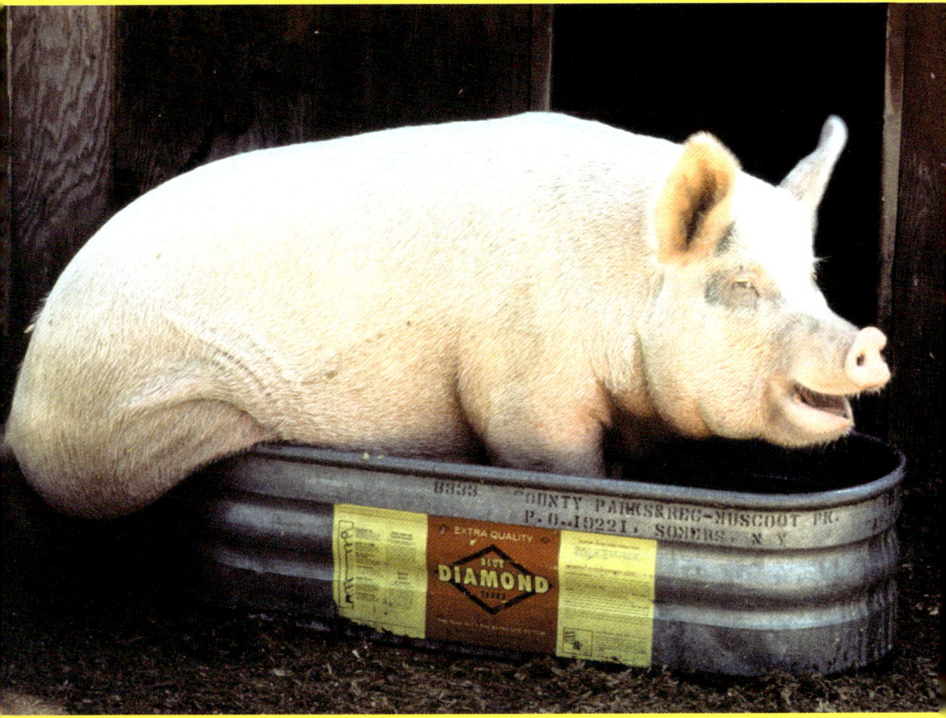
EXTRA QUALITY
DIAMOND

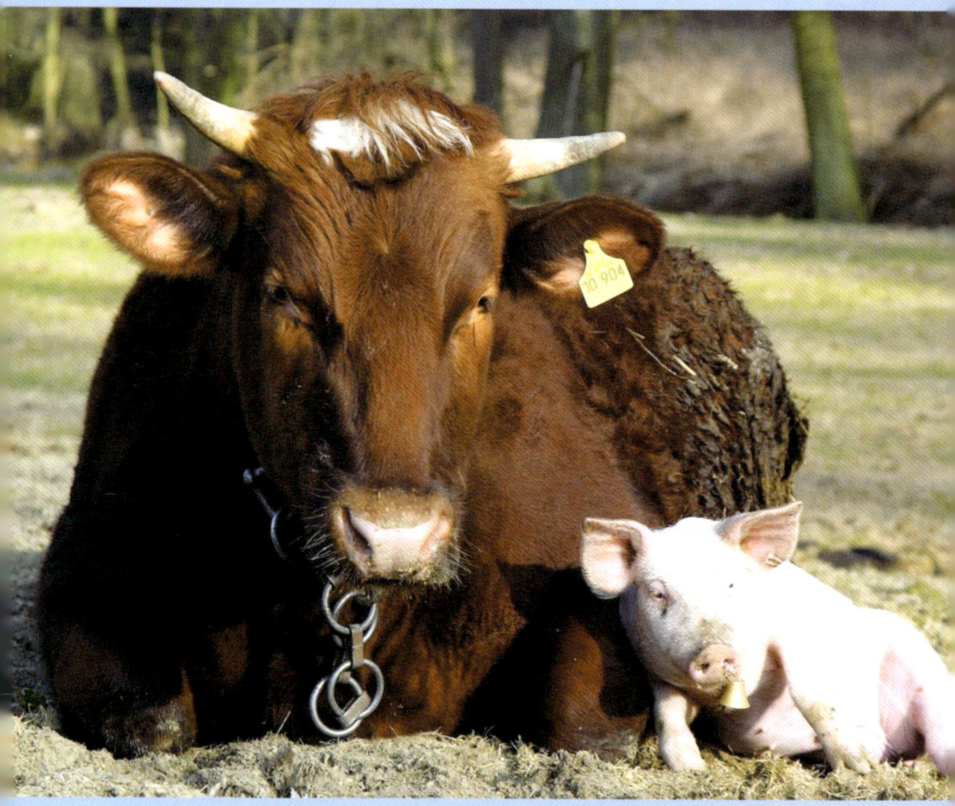
10 904

'Oh, The Farmer and The Cowpoke Can Be Friends'

Billy Bull and Peter Piglet follow their masters' lead.

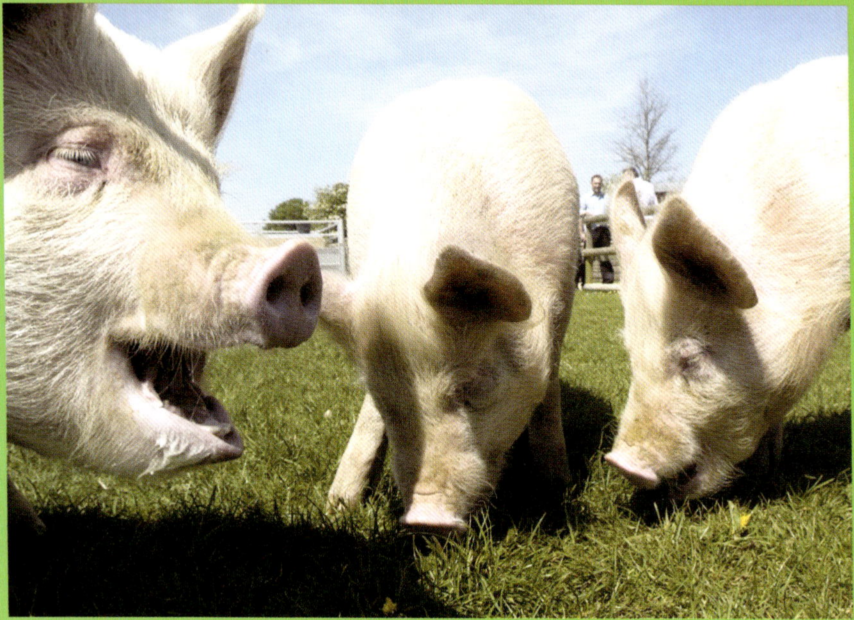

Huff Daddy and the Puffs

'I'll sing it, you ham it.'

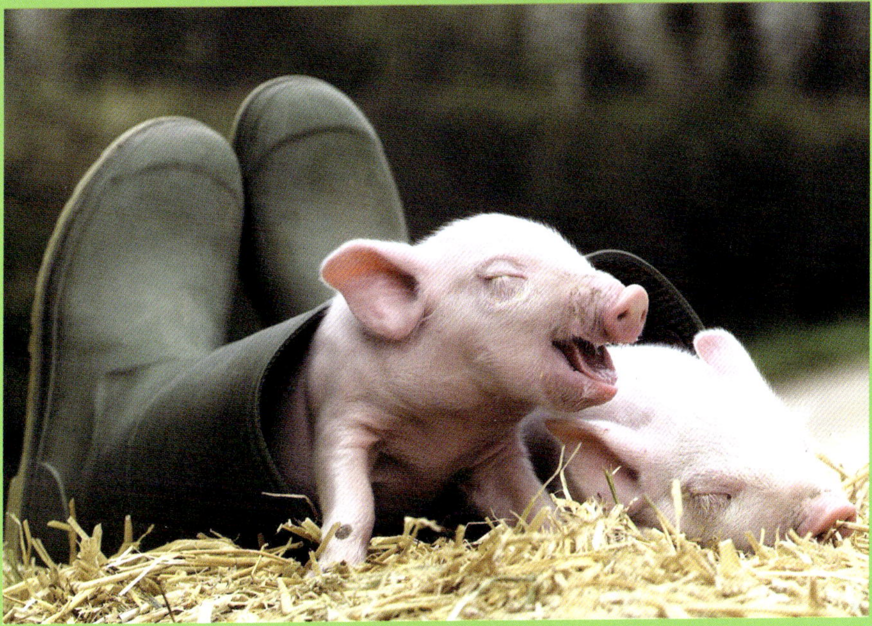

These Boots Are Made For Waking

Grumpy gives Dopey a wake-up call.

Inflated Ego

The piglets give their new stall-mate
the seal of approval.

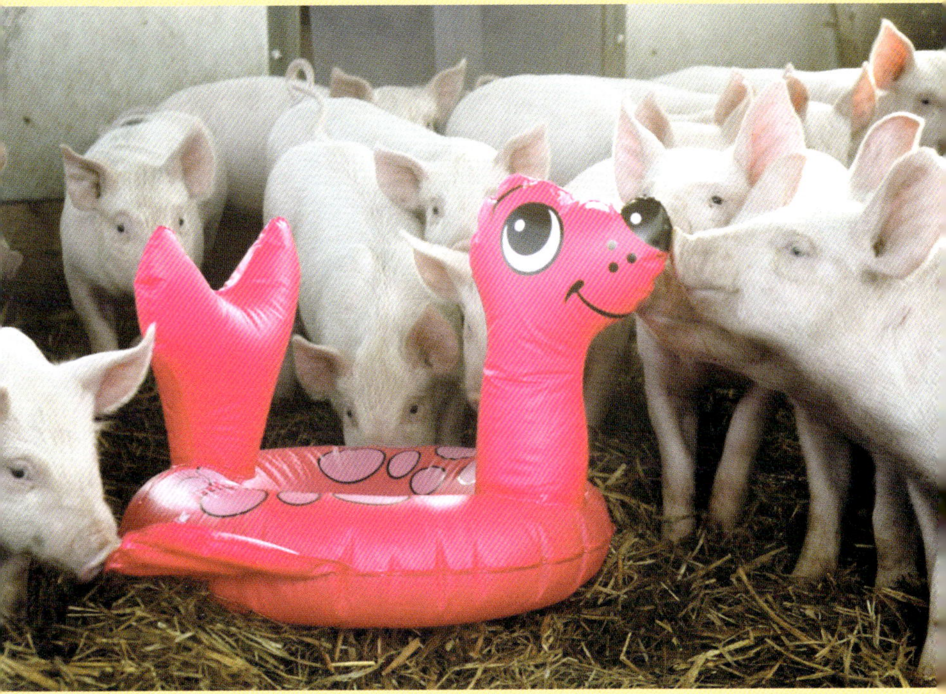

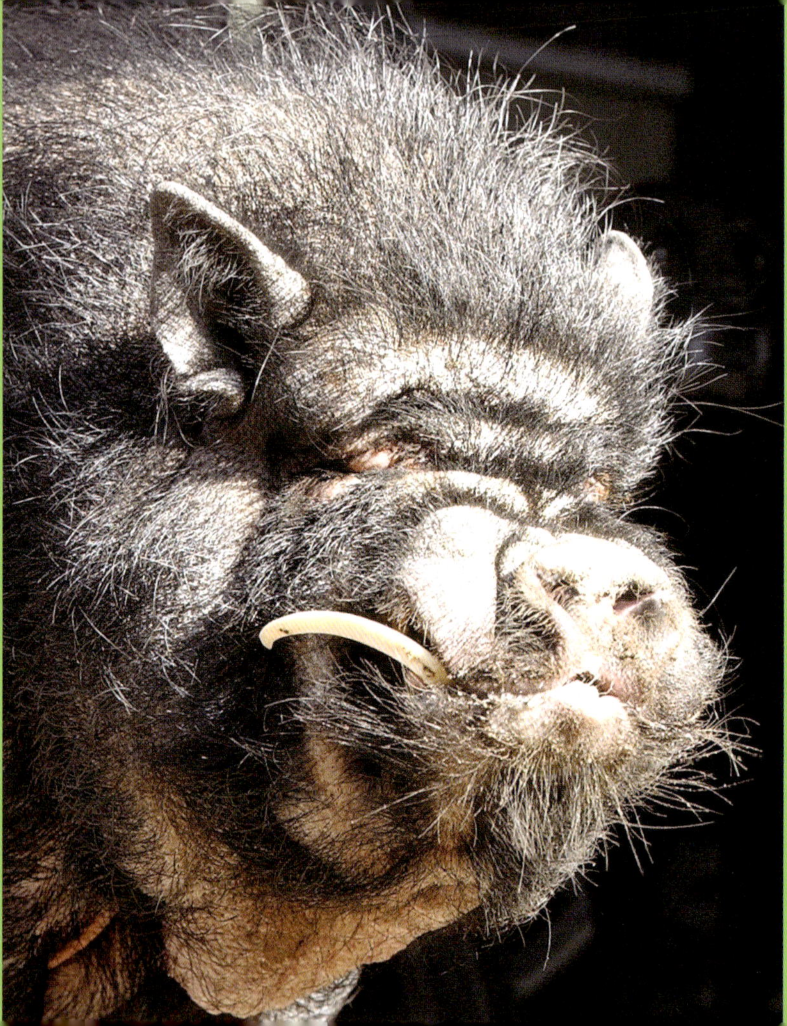

Raw-Hyde

Jekyll experiences an unexpected moment of boar-dom.

Pork the Pork

'Gammon over to my place!'

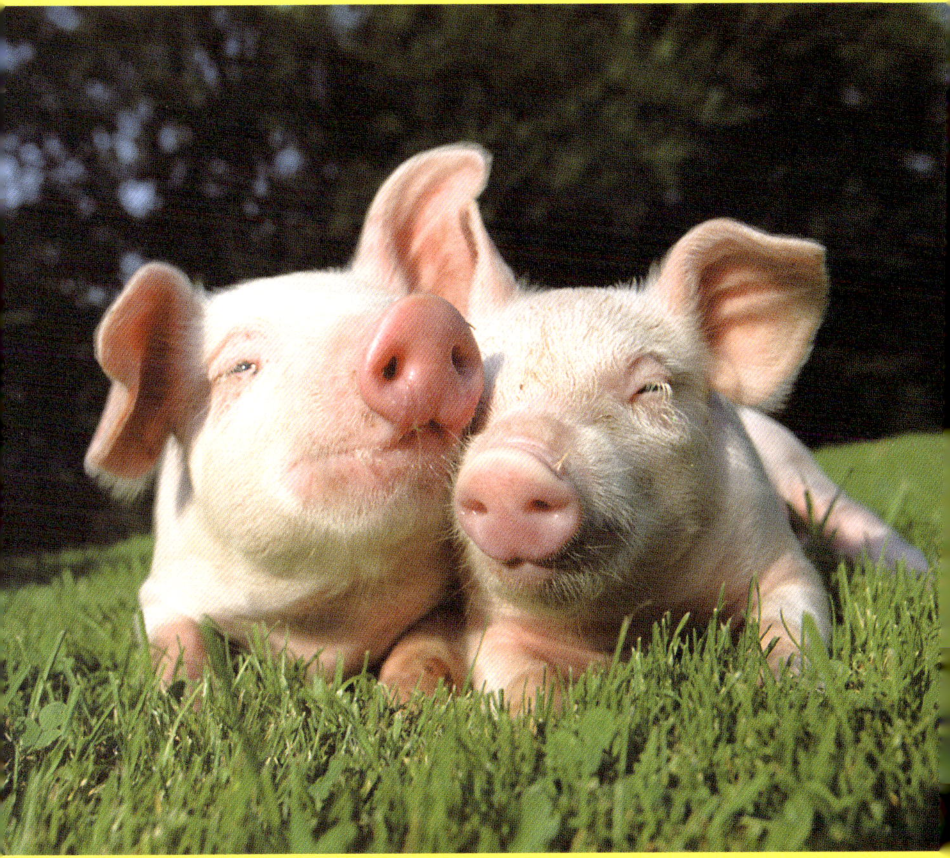

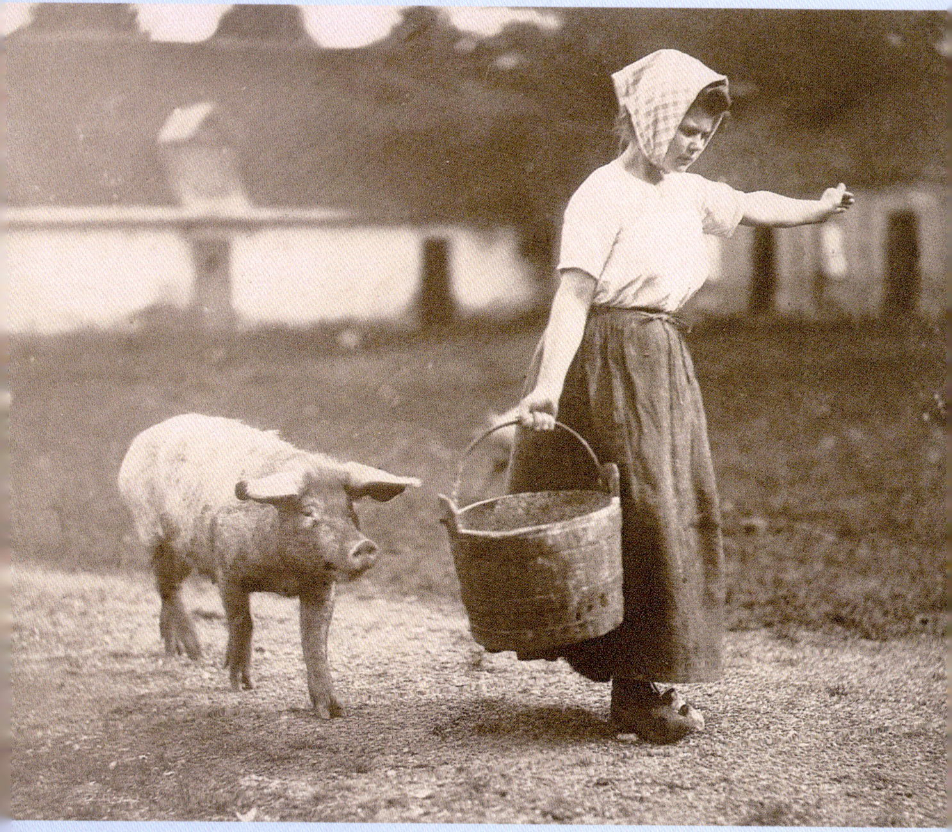

Follow My Leader

Hyacinth stays close to the bucket as you never know where the next meal is coming from.

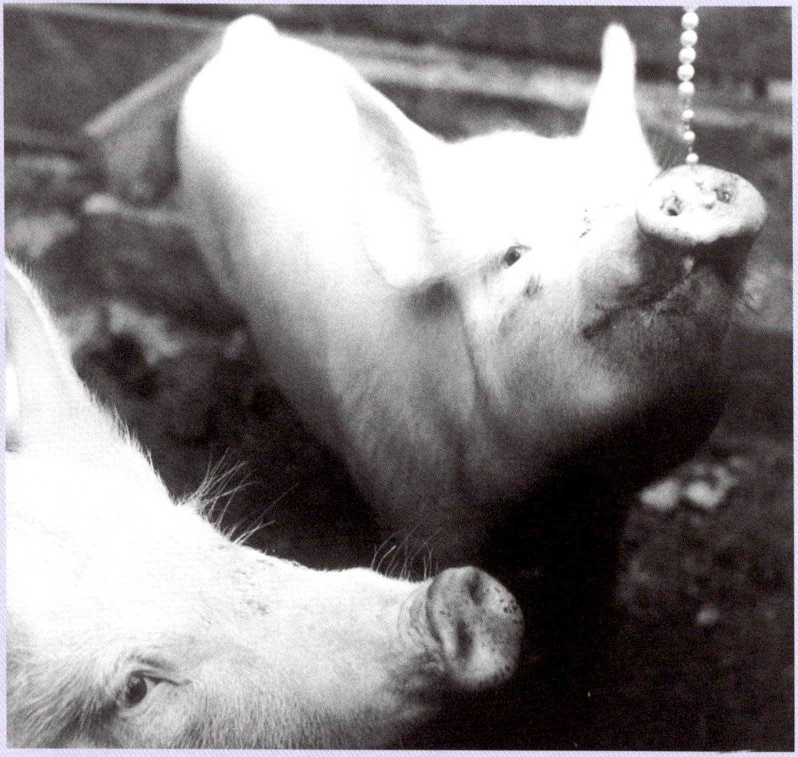

Pearls Before Swine

Victoria still thinks diamonds are a girl's best friend.

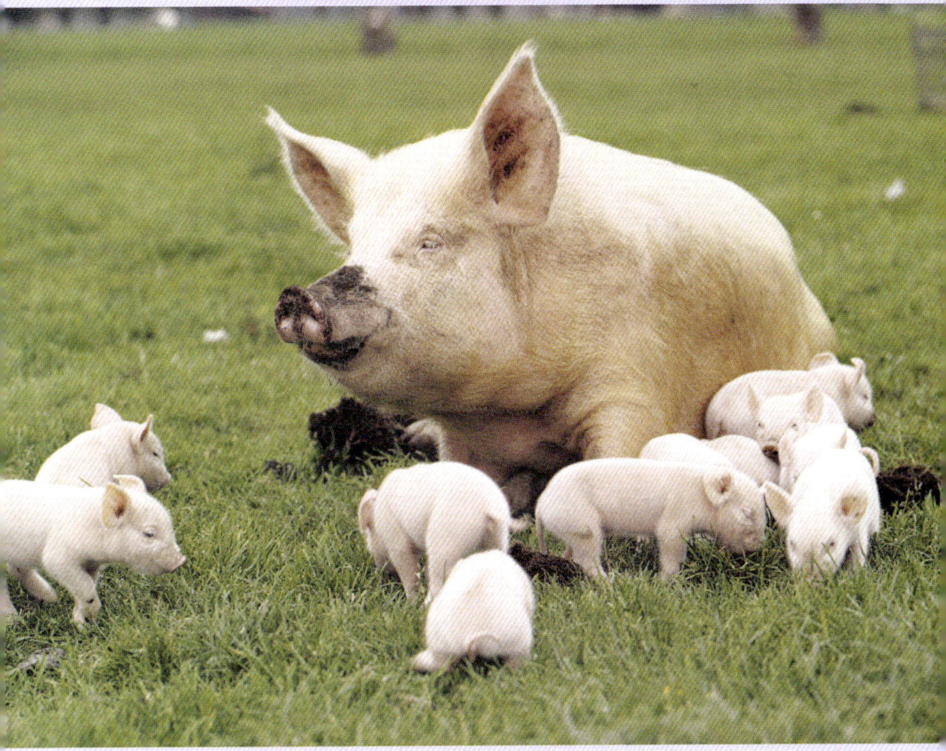

Motherly Love

Ma looks on fondly as her children make pigs of themselves.

For My Next Trick . . .

Cuthbert wonders if juggling was the right career choice for him.

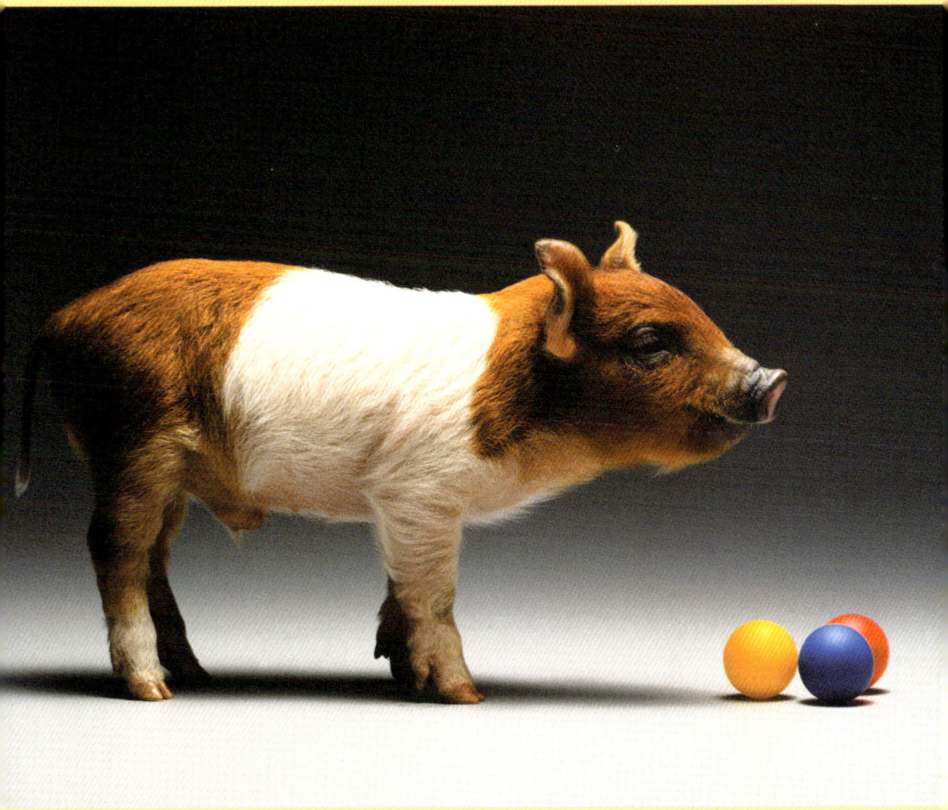

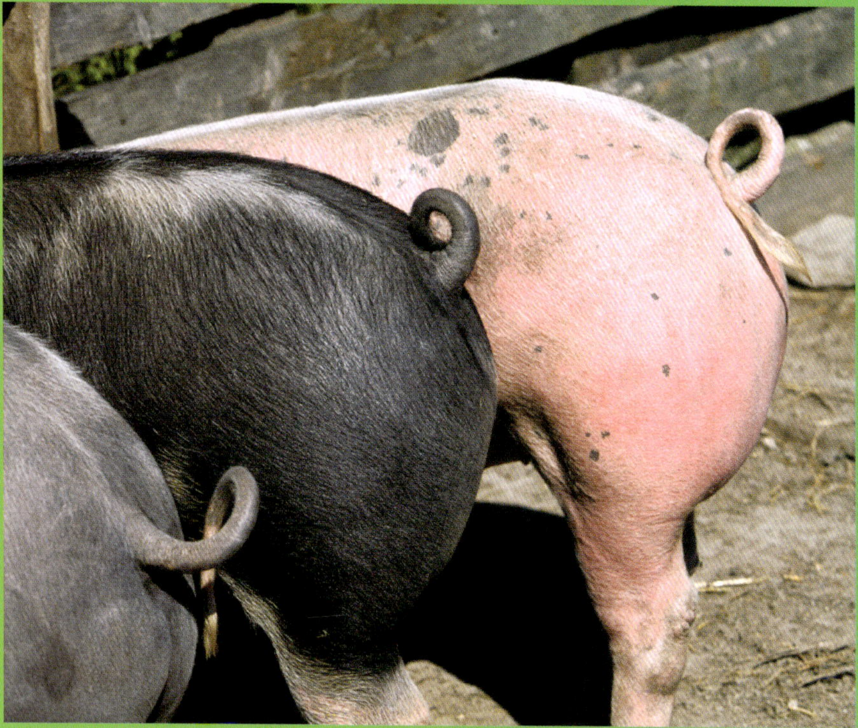

Synchronized Swine

Three contestants practise for the final.

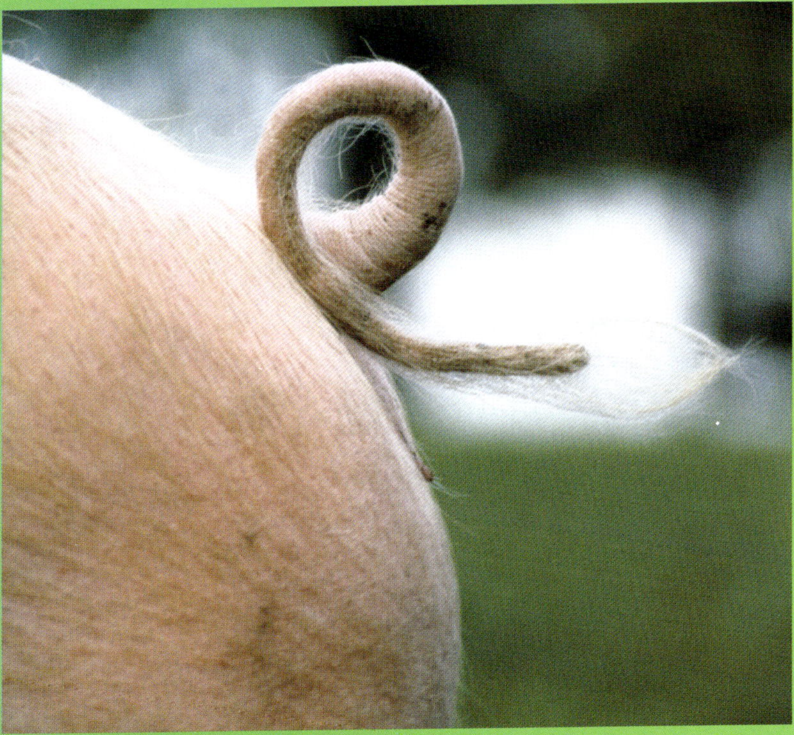

In A-Rears

Hereby hangs a tail . . .

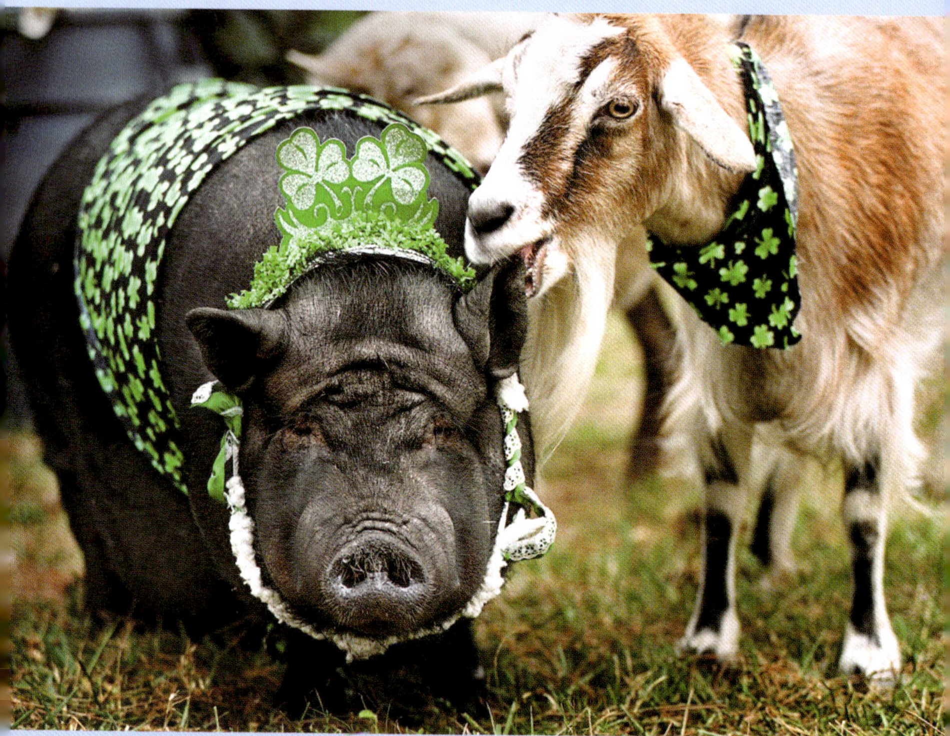

Saint Paddy Pairing

Mildred kissed the Blarney stone and starts whispering sweet nothings in Seamus's ear.

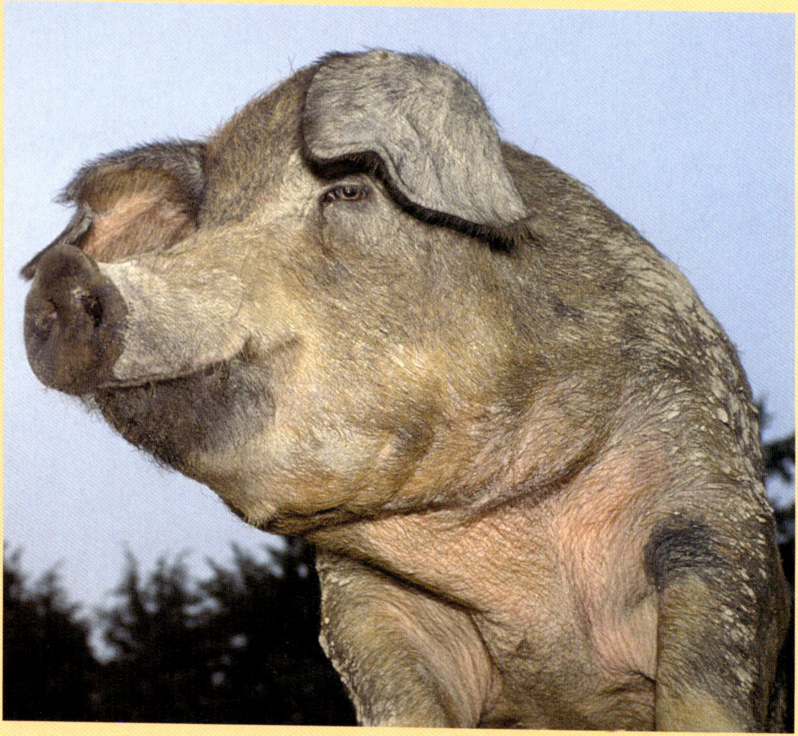

Hogging the Limelight

'Okay, Mr de Miller, I'm ready for my close-up now...'

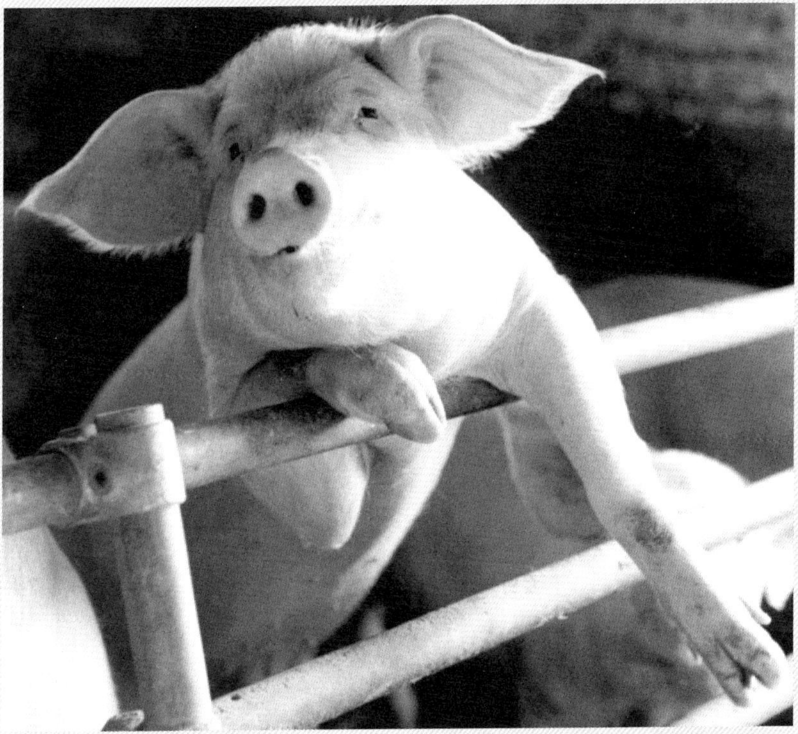

Lend Me Your Ears

Julius delivers a lecture on the pig bang theory.

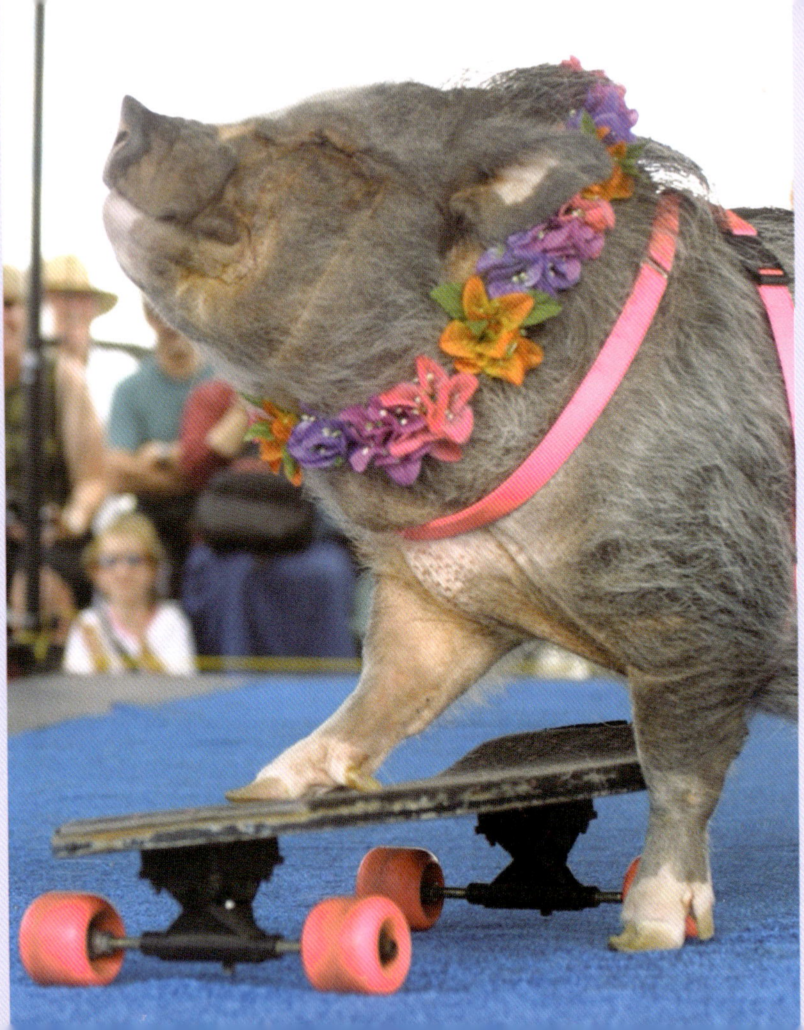

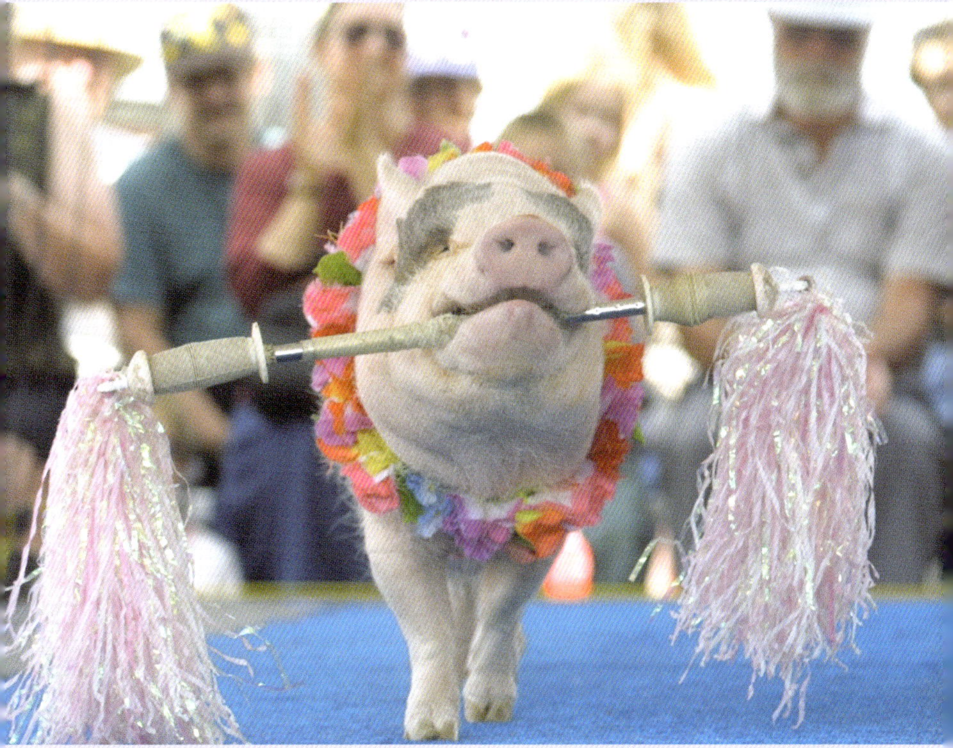

On the Board

Wilbur shows his wheely wild side.

Get Your Ra-Ras Out

Priscilla is the chief cheerleader for the New Pork Vets.

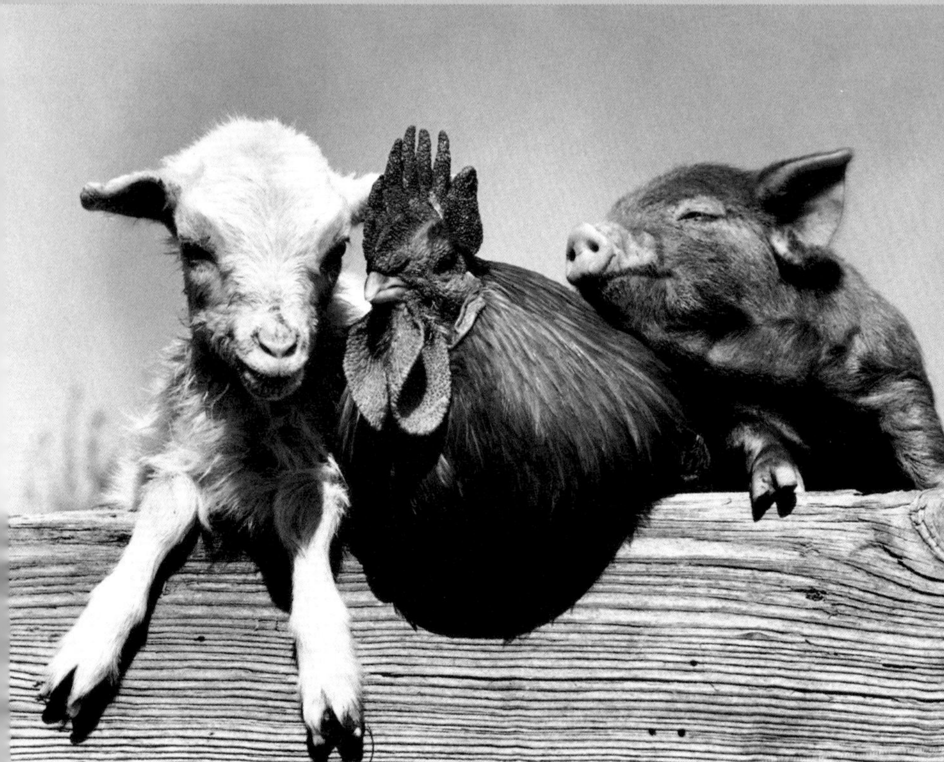

Don't Fence Me In

Three little buddies become farm friends.

Being Nosy

Bessie wants to know where Betty and her friends got their fancy coats.

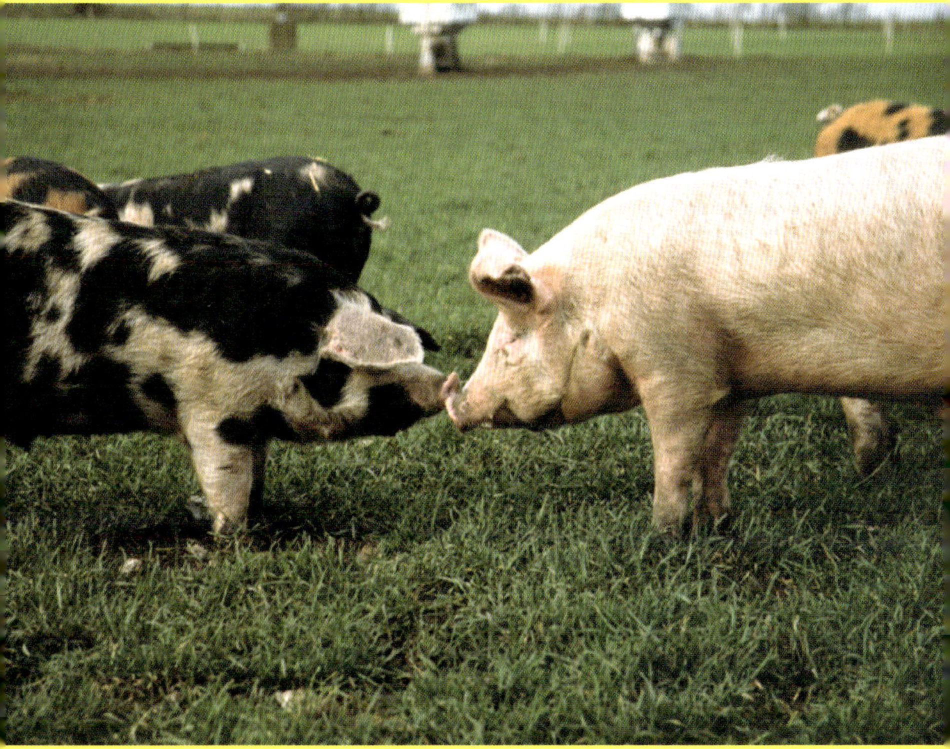

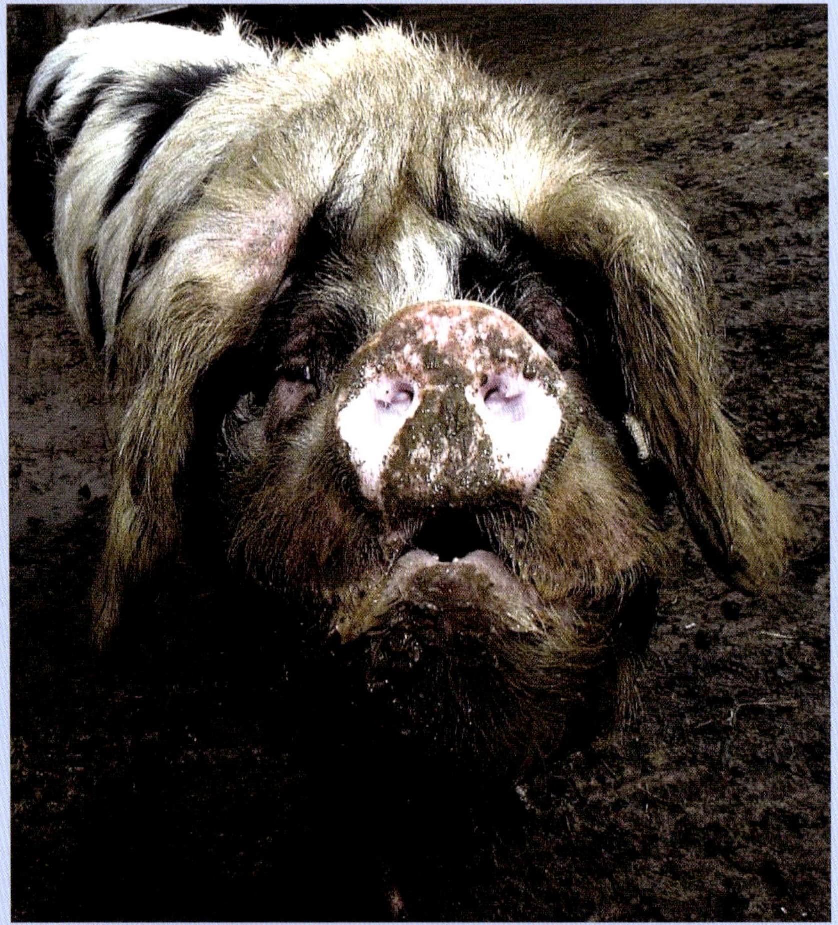

Pig Ugly

Henrietta has a bad hair day.

Every Which Way

Ma moans that she doesn't know if the kids are coming or going.

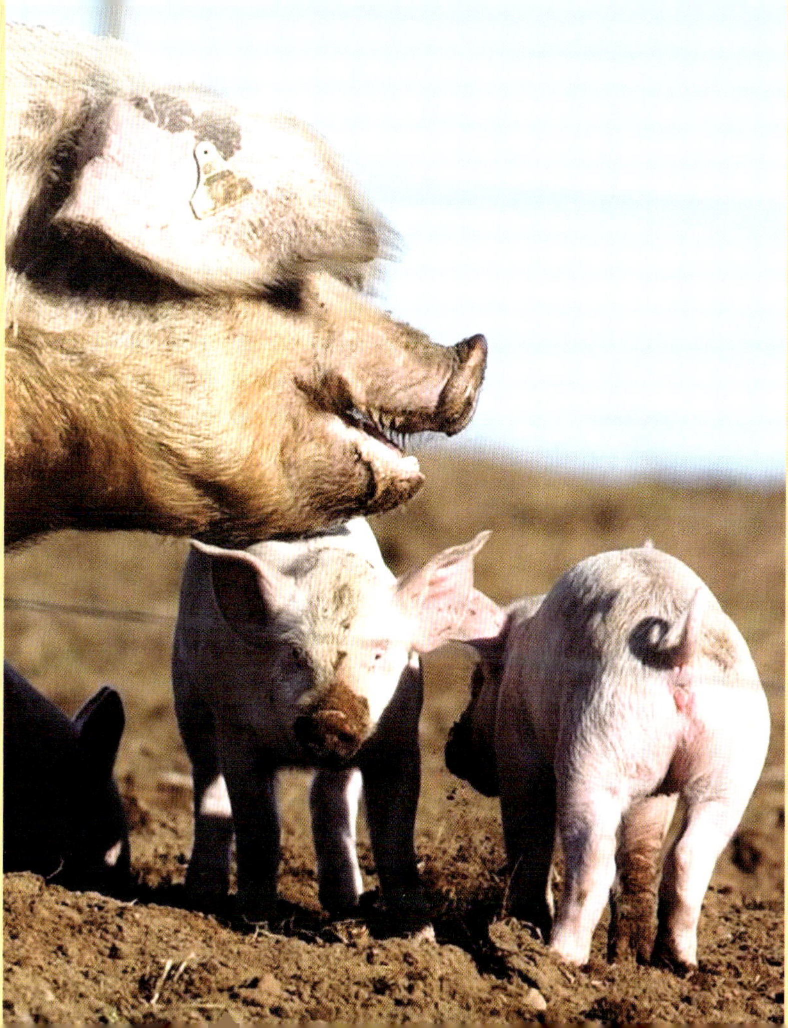

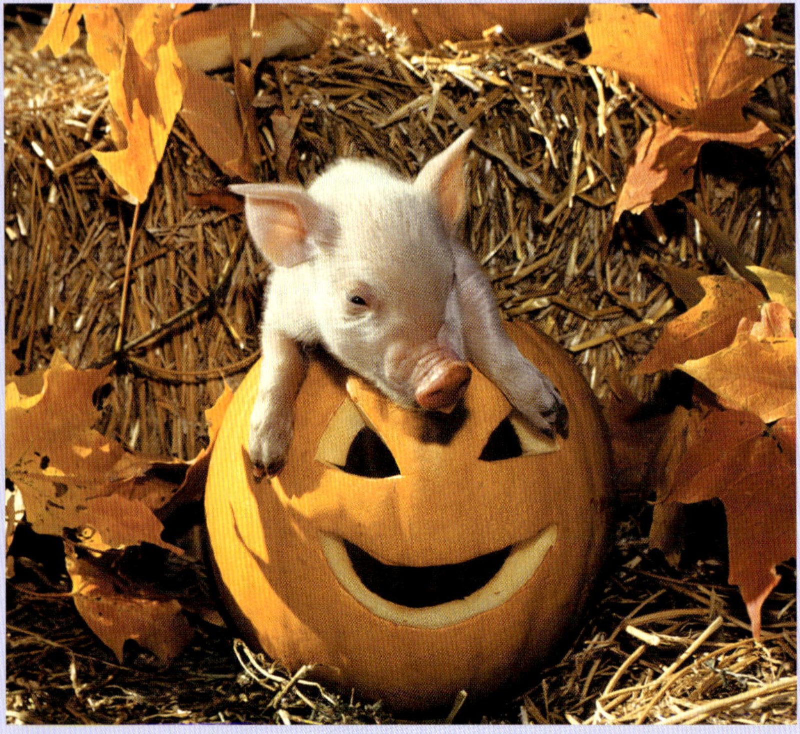

Halloween Ball

Jack O'Lantern is the light of Little Pumpkin's life.

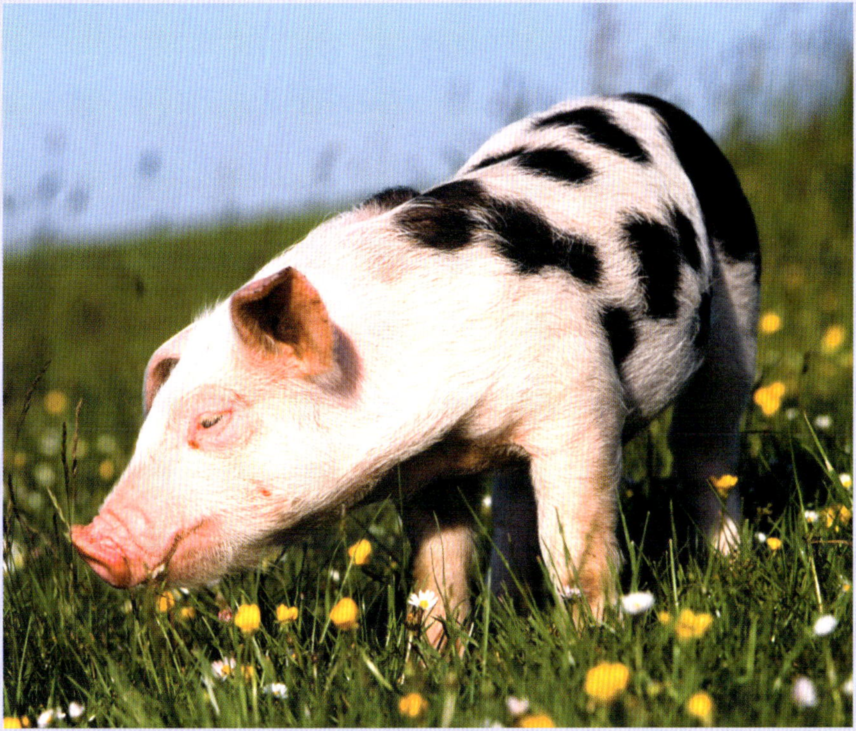

Sniffer Hog

Horace hopes for rich piggings.

Dullas Ditchwater

JR thought he was going to be starring in a new soap opera.

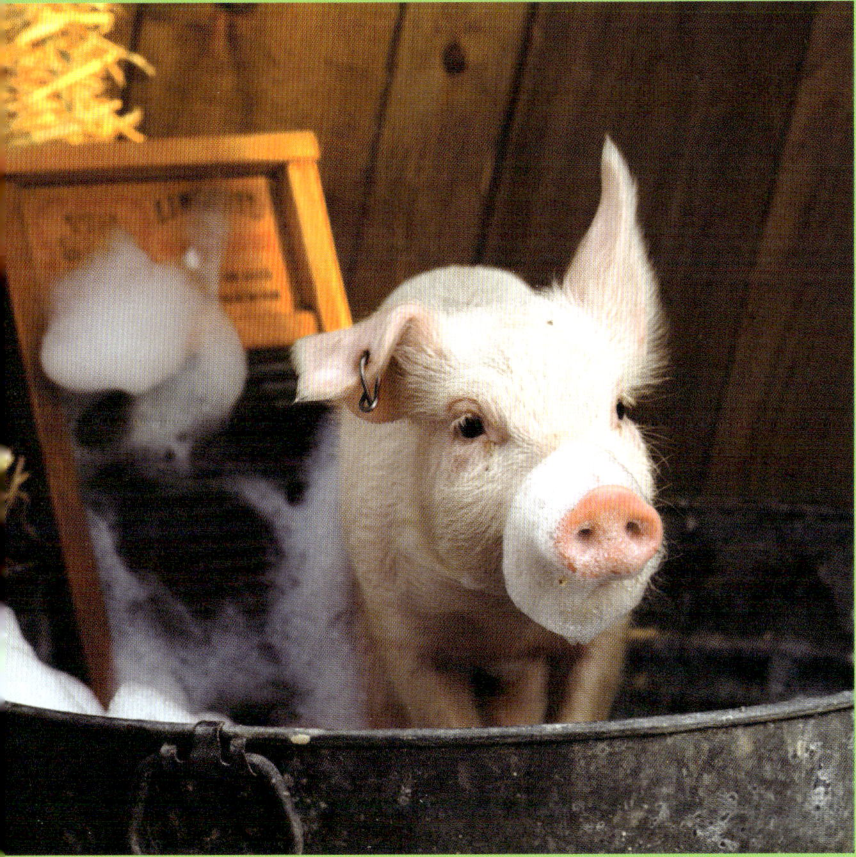

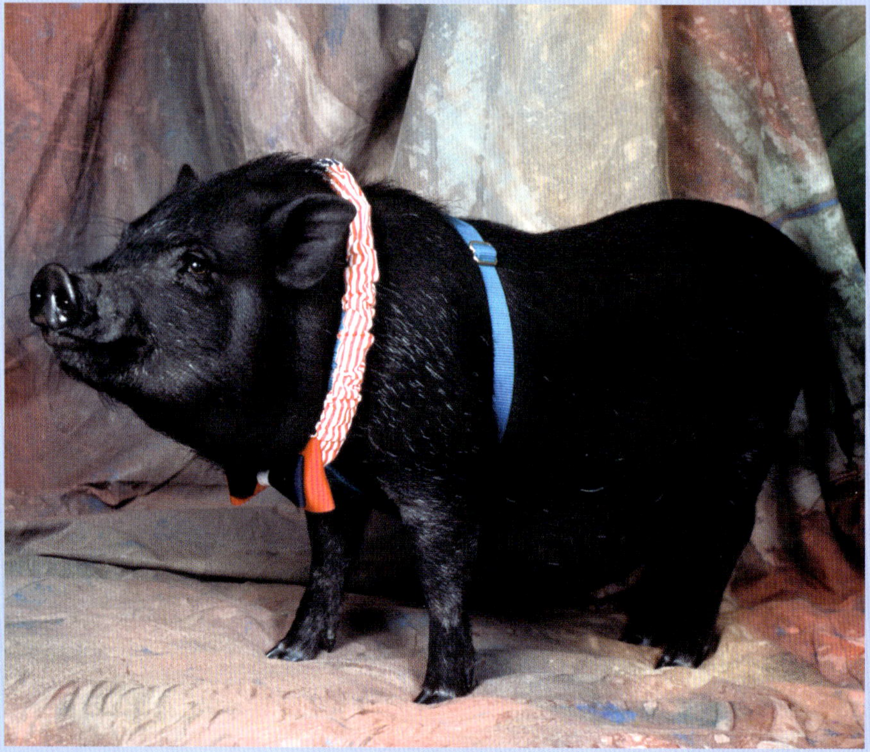

Prize Pupil

Hegridd graduates with honours from the Hogworms School of Piggery.

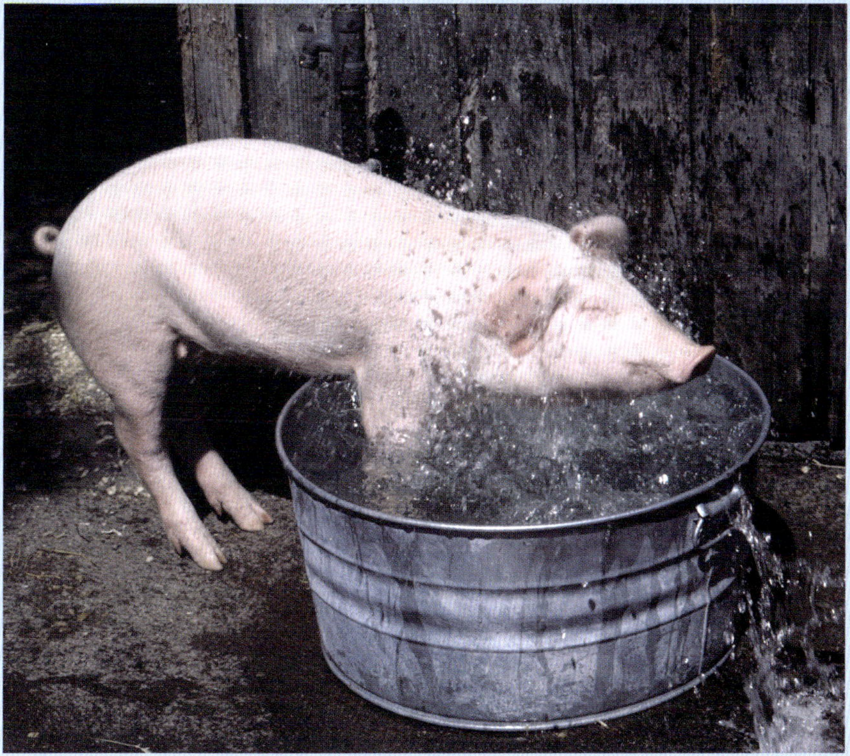

Hogwash

Brutus splashes it all over.

Porky Winks

To sleep, pork-chance to dream...

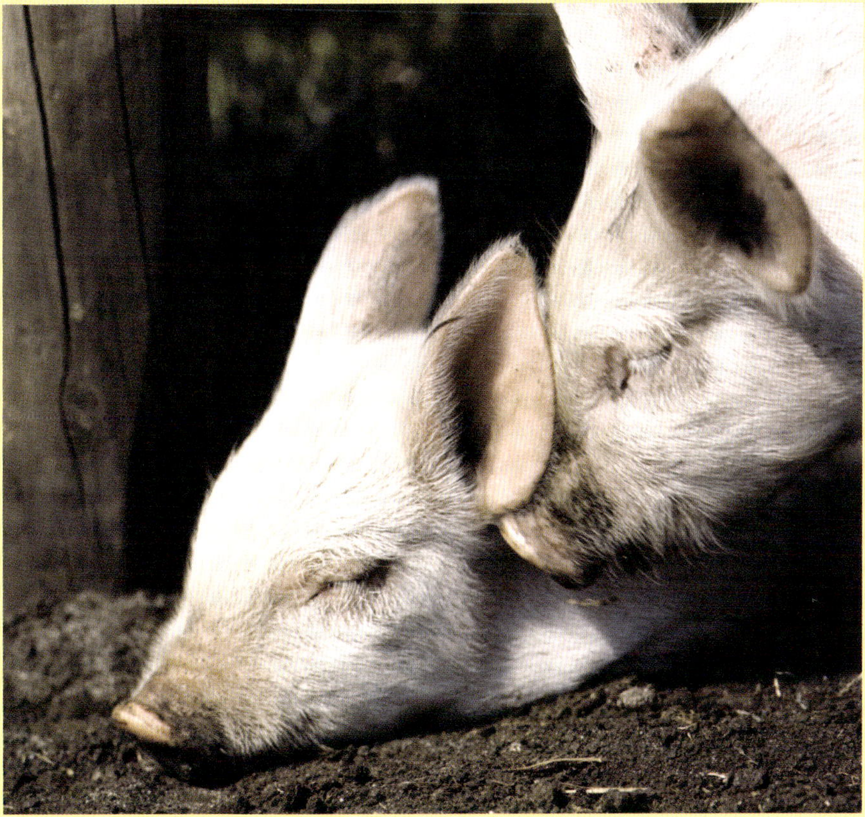

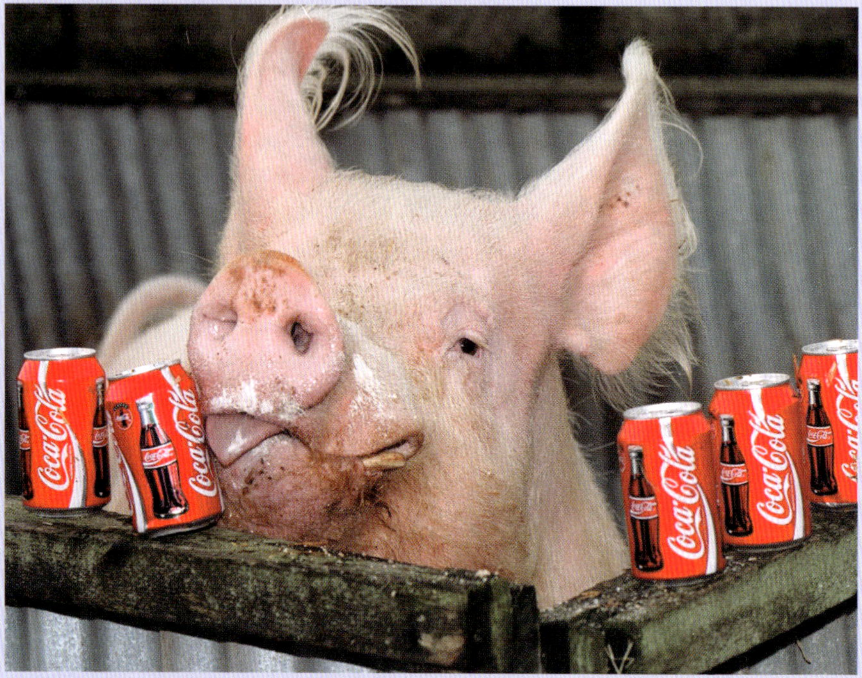

Happy Hour

Larry lines them up.

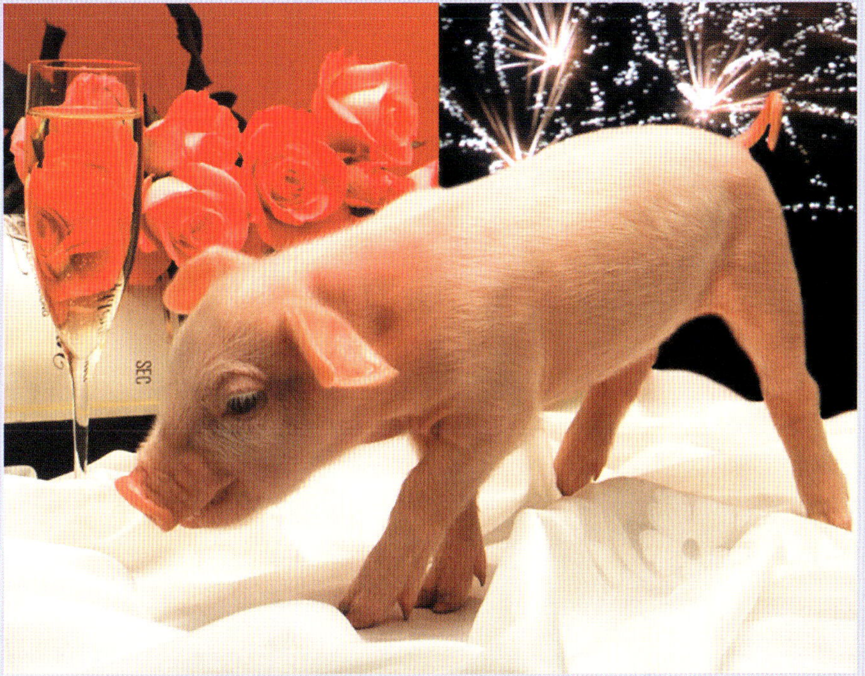

Days of Swine and Roses

Henry goes the whole hog for Valentine's Day.

Stretch-Limo

Clara calls for a pink Cadillac to take her to town.

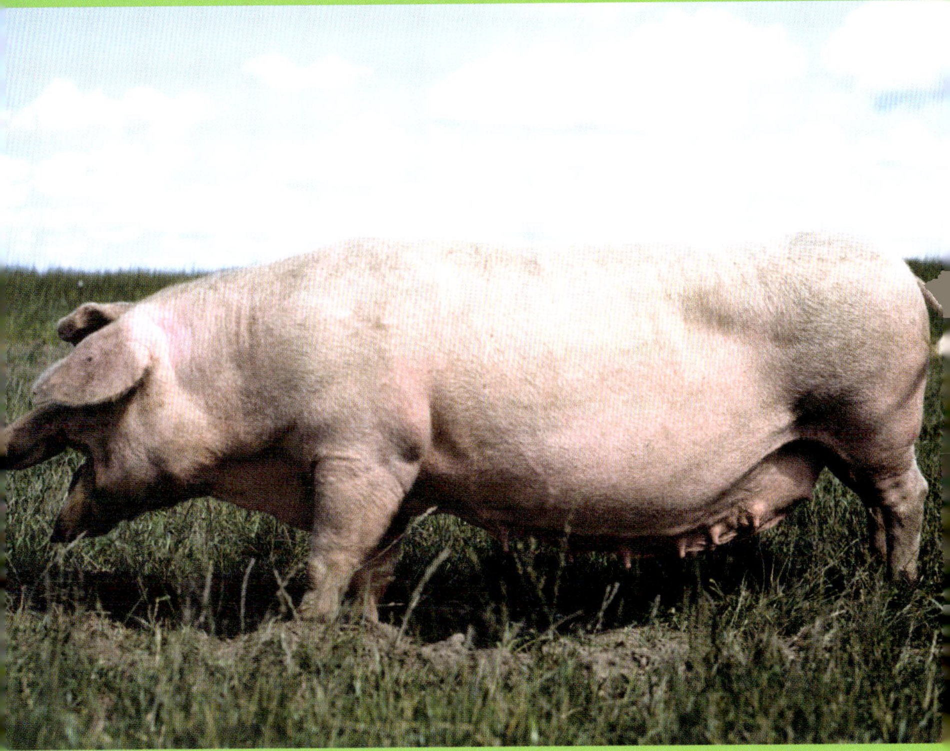

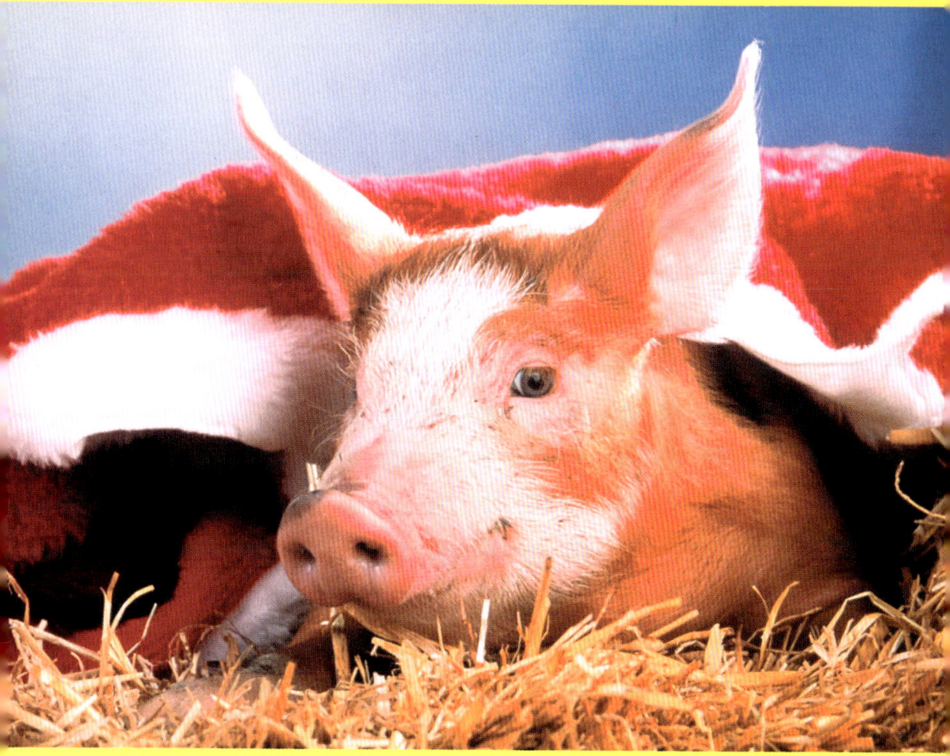

Santa Baby

Too excited to sleep on Christmas Eve,
Peggy listens for sleigh bells.

Boarseye

Tel's not sure he trusts William with that arrow.

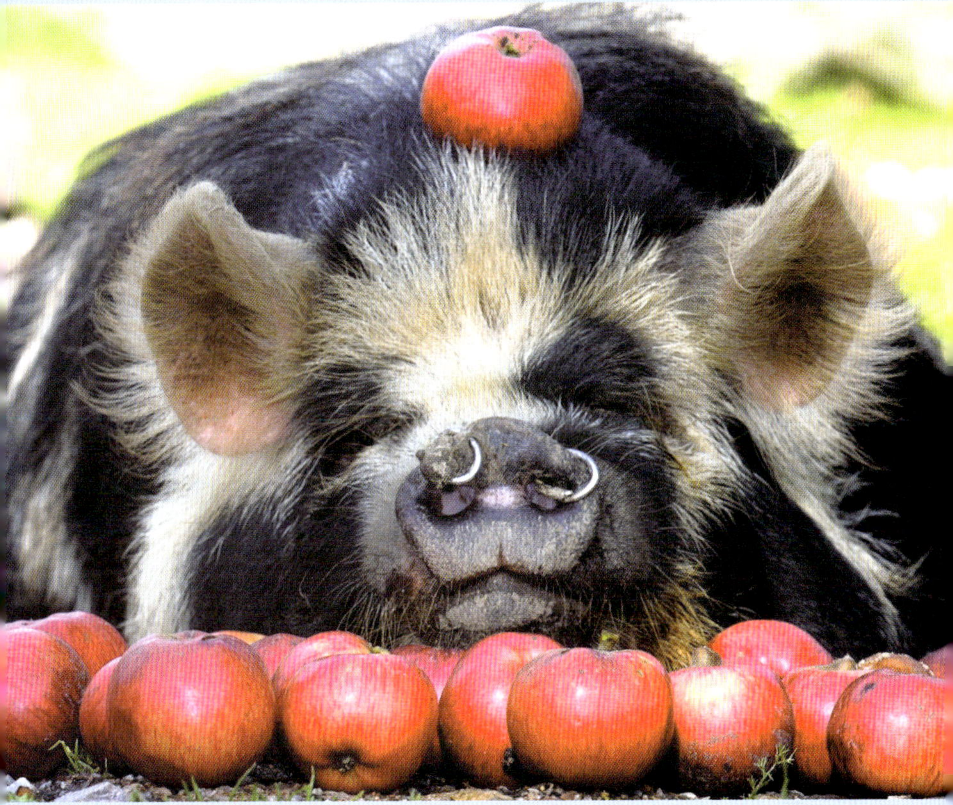

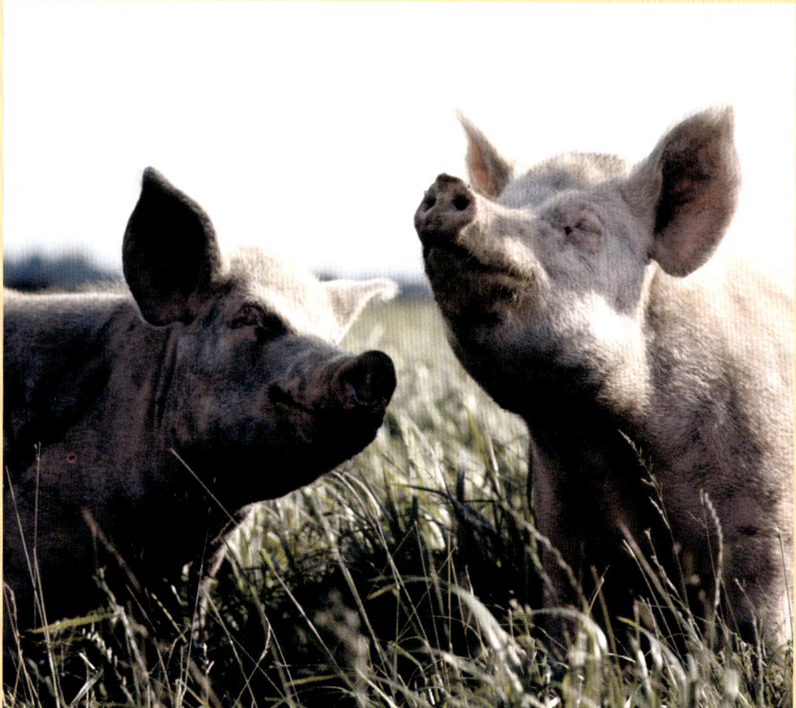

Scent Off

Sam and Samantha savour the sweet smell of success.

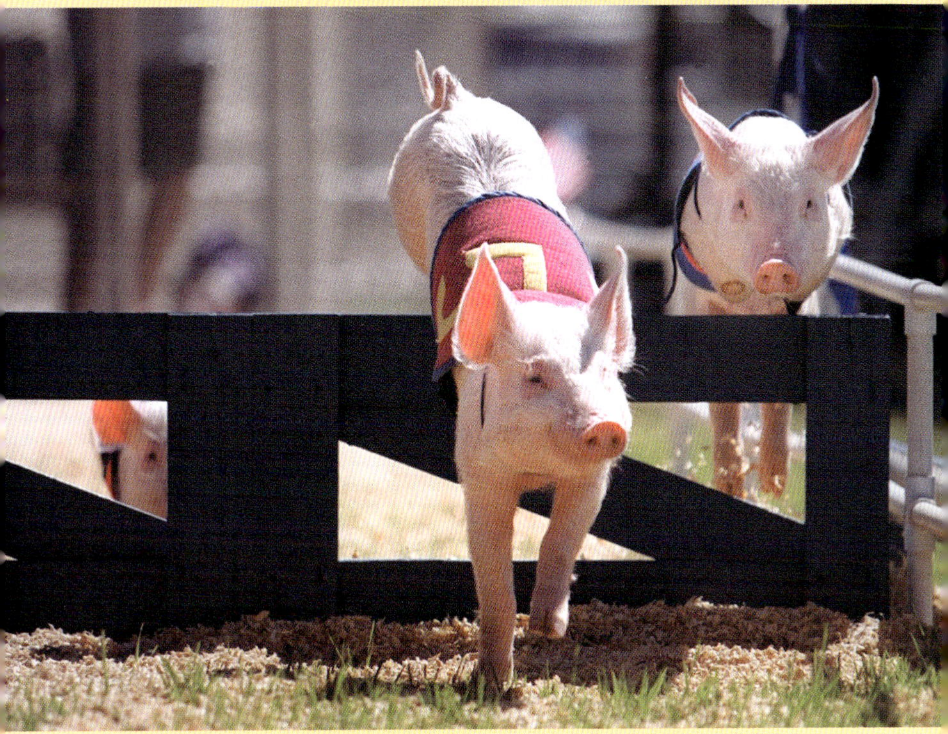

A Day At The Races

Hop, Skip and Jump take all obstacles in their stride.

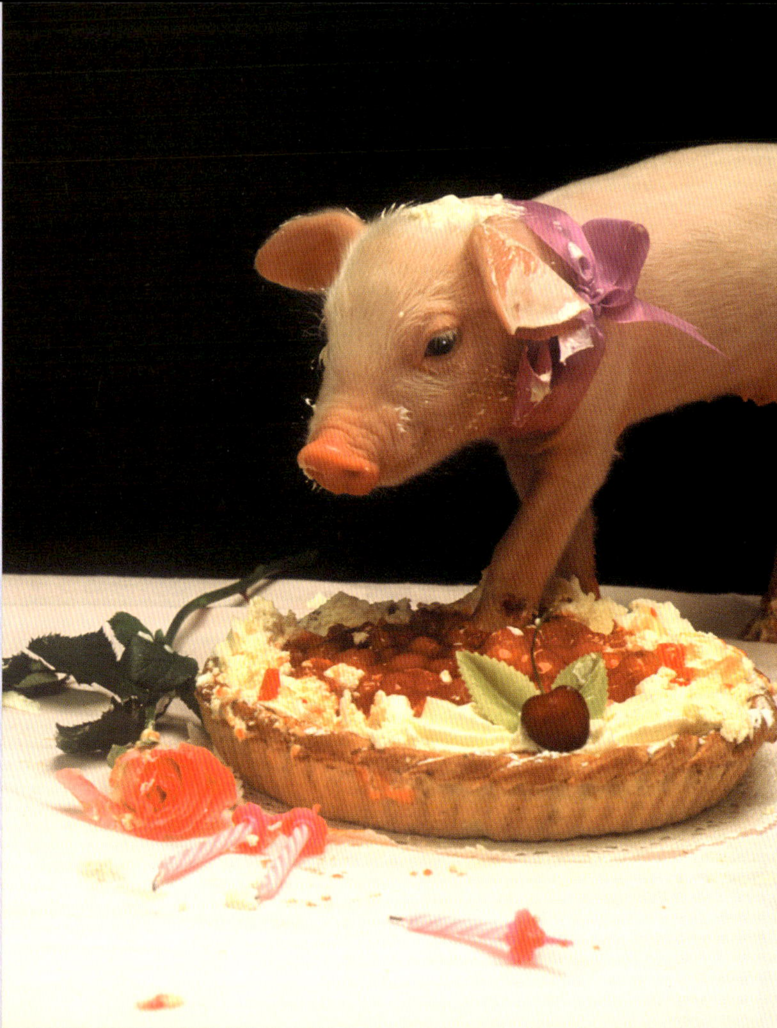

Pie in the Sty

Harry Trotter puts his hoof in it.

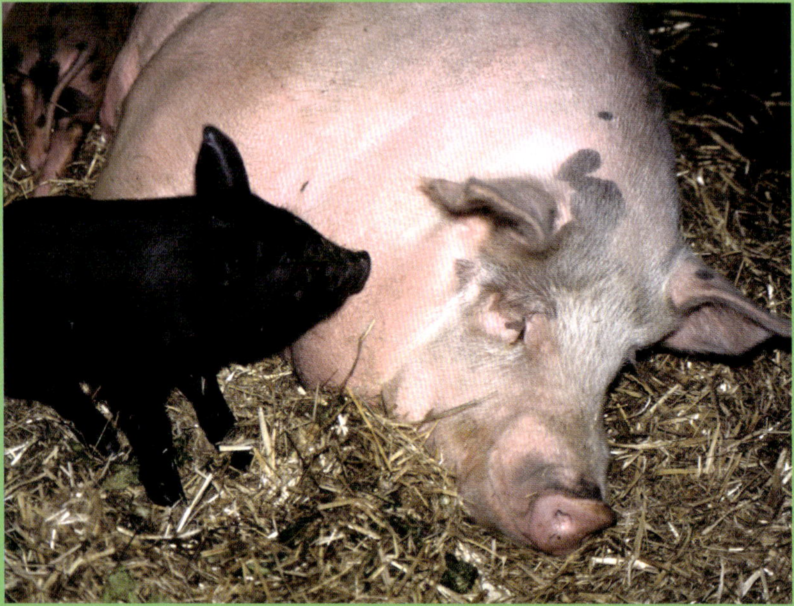

Black Looks

Sooty is certainly not a chop off the old block.

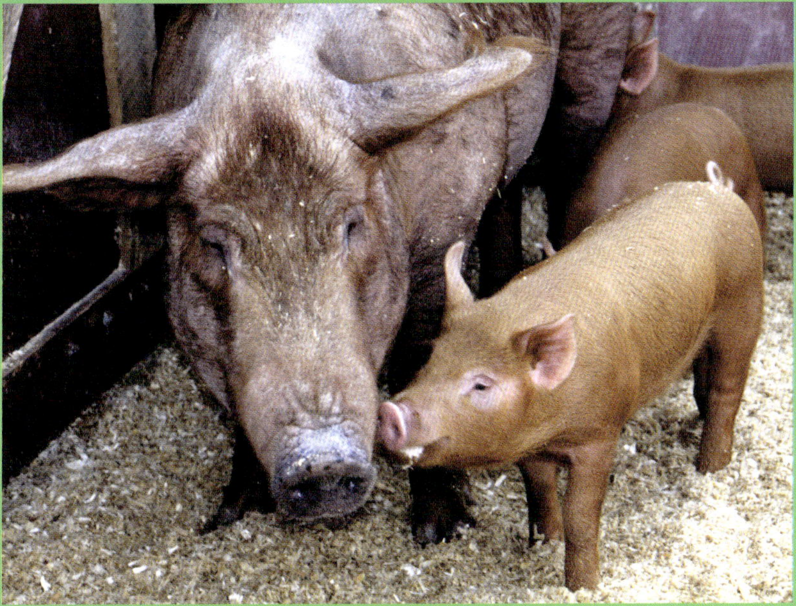

Pig-ment of His Imagination

Horace askes his uncle if he is one of the ‘pigs who might fly’.

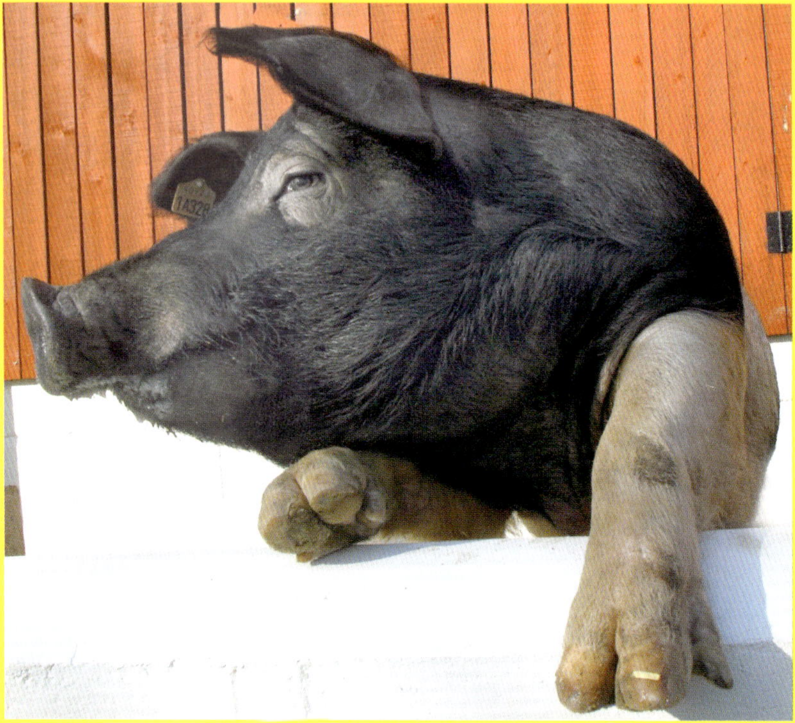

Pig Brother

Some pigs are more equal than others.

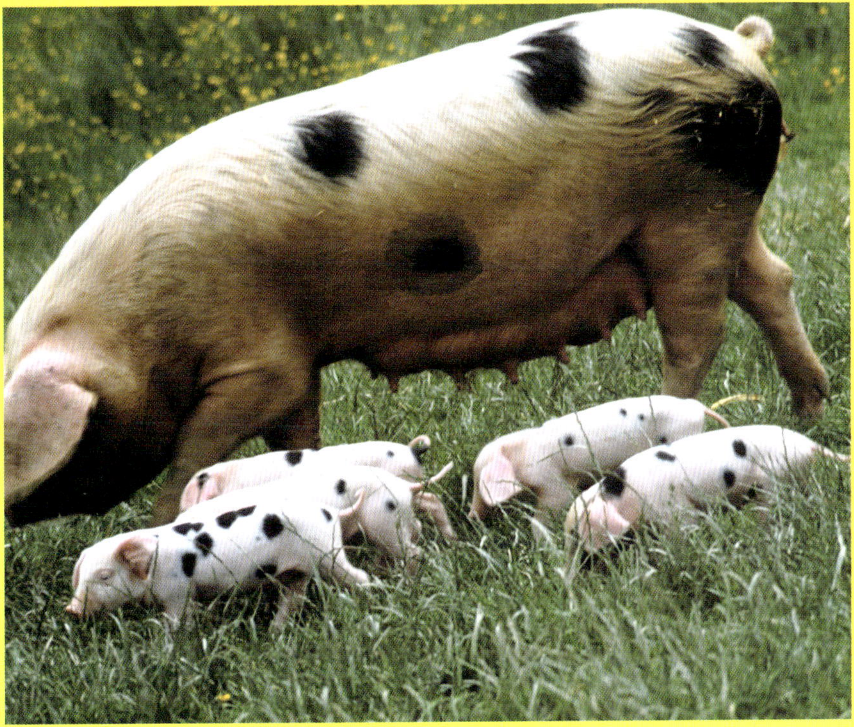

Spot the Difference

Dotty finds it difficult to tell her darlings apart.

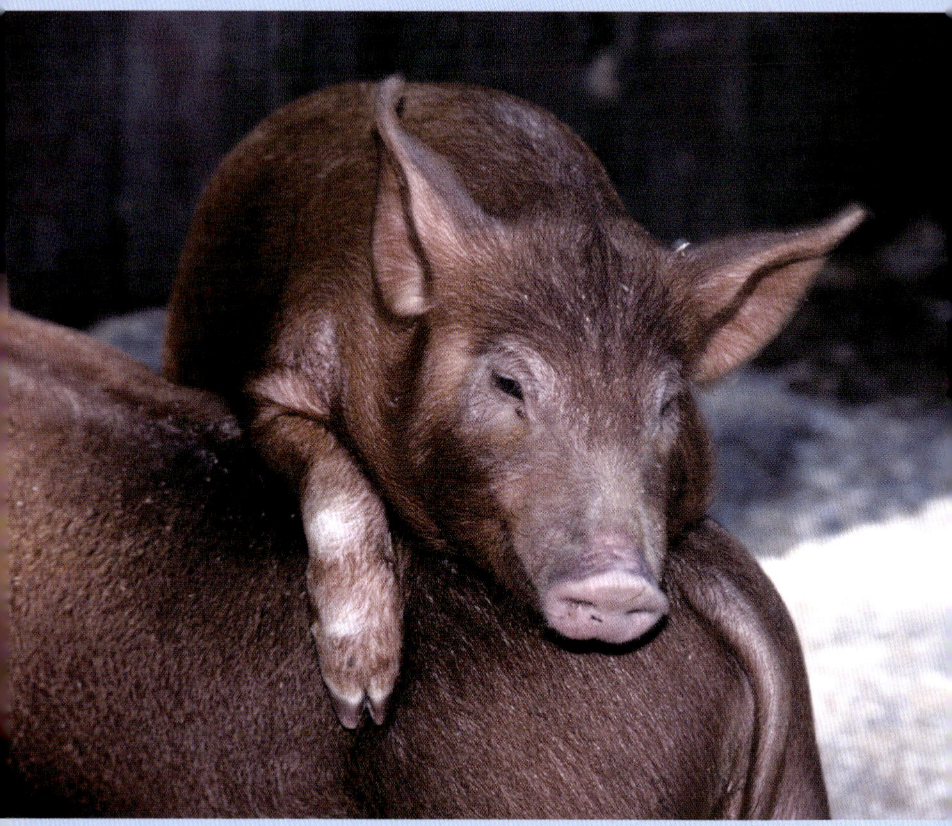

Piggy Back

Hal hitches a ride.

Getting In A Flap

Rory makes a fast exit!

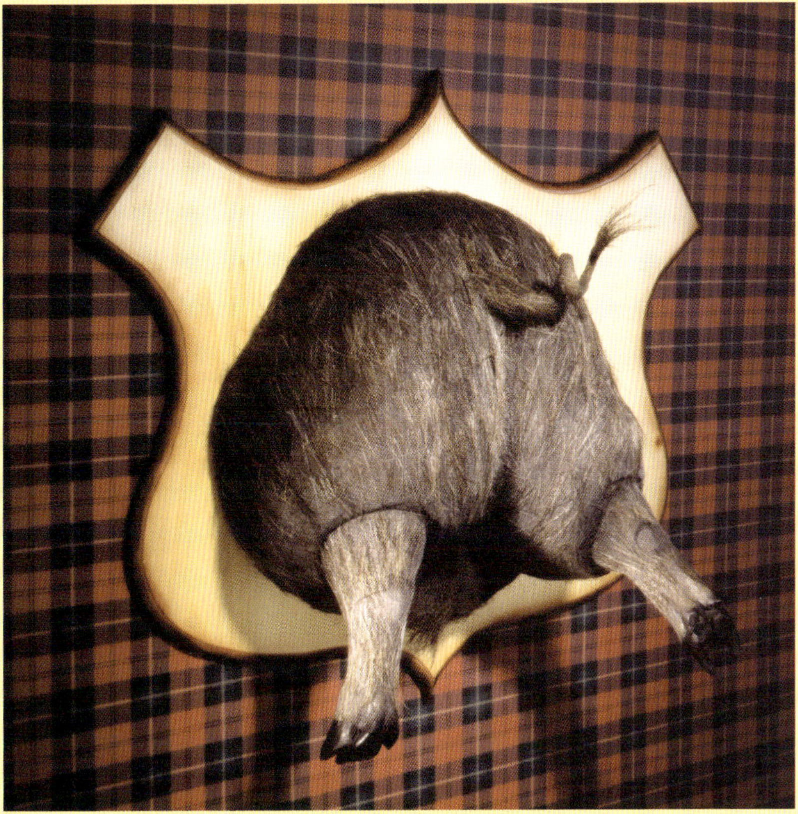

PHOTO CREDITS

Alamy
ImageState 12, 15; Gay Bumgarner 40, 52; David Hoffman Photo Library 47; PhotoValley 61 (Kris Coppieters), 70

Corbis
13

Corbis at Cadmium Systems Ltd
4, 35

Getty Images / Hulton Archive
8, 11

Matton Images
Image100 9, ImageState 55

Powerstock
19, 29, 32, 33, 36, 37, 44, 53, 56, 57, 59, 63, 68, 72, 73, 76, 79

Rex Features
Simon Webster 16; Action Press 20; Nils Jorgensen 22; Richard Austin 23, 24, 67; Mike Walker 26; Jennifer Podis 38; Les Wilson 48, Northscot Press Agency 50; NWI 60; U. Schanz / Fotex 64; Gary Roberts 74

TopFoto.co.uk
30, 41, 42, 43, 69, 75